Advanced College Algebra Study Guide

An engineering approach to learning, understanding, and teaching Algebra

A visual, procedural and verbal, learning guide.

Mr. Cook
Principal Electrical Engineer
Adjunct Math Professor
October 31, 2023[1]

Copyright Page

Library of Congress Control Number: 2020921703

PCN: ADVANCED COLLEGE ALGEBRA STUDY GUIDE

PCN: 9781735463520.

ISBN: 978-1-7354635-2-0 Advanced College Algebra Study Guide

Worcester Massachusetts

First edition Advanced Algebra Study Guide

Second edition College Advanced Algebra Study Guide

Second Edition Changes:Title change and rewrite of first edition

©2020 Harrison Cook

Appreciation

My Family

Susan S., Library Consultant

Brandon, Math Student, FSU

Contents

List of Figures

Prefix, Introduction

After a career in Electrical and Computer Test engineering, then teaching math in college I compiled much of my notes used in the classroom to help future students gain math instinctive problem-solving skills, using the tools presented in class.

I decided to write a study guide after teaching Algebra for 12 years in a local community college. Both Electrical Engineering and Practical Math were used every day of my projects. It was our task to debug new designs and design test equipment and make both the designs and test equipment work before the design could be approved. We used a lot of practical mathematics putting to use the Math and physics and electrical theory that we learned in engineering school. When I switched to teaching the problem was how to teach math so it could be understood.by students not necessarily comfortable with the Math they were taught in High School. To that end I started writing proofs that were used as handouts to the students.

Methods ,Graphs and Proofs

Discussion of the basic principles and in most cases a diagram or graph to accompany the discussion.

- Look at underlying rules and principles

- Graph when needed and view the math in 2-dimension or section of a curve.

- Compare with Algebra and Geometry.

Look at underlying rules and principles

Graph when needed and view the math in 2-dimension or section of a curve.

Compare math formulas with Algebra and Geometry.

0.1 Scope of this Book

The Scope of this book is to provide a

study guide for many of the principles covered in my

College Algebra Courses.

0.2 Results of the study guide

The result of this study is intended to give the student a better

understanding of a solid approach to solving math Problems. to

In subsequent tasks in the Engineering field the

engineer may need to re-derive the principle or equation to solve the problem presented by the required task.

The engineer uses underlying principles of math to create mathematical

flexibility to solve problems in the class room or engineering tasks in the workplace.

For example my task was to determine if a board tester could test CMOS integrated circuit without destroying the printed circuit board. The results were mixed on the capibilities of the tester and the tester had to be replaced eventually. This was justified by measurements and calculation to demonstrate the electrical characteristics of the board tester.

Please feel free to check out my website:

mathstudyguide.org

Click on the picture of the book to be directed to Amazon to purchase the book.

Chapter 1

Real Numbers

The following diagram is an example of the types of real number lines in the real number system. Note that the rational number system consists of all the numbers on the real number line. In a graph of the real number line this would represent the x-axis. The cartesian coordinate system consists of the real number line plotted on the x-axis and the y axis. The y axis is perpendicular the the x-axis. The two lines intersect at x,ycoordinates at point 0,0 The diagram is an example of the types of real number lines in the real number system. Note that rational number system consists of all the numbers on the real number line. In a graph of the real number line this would represent the x-axis. The cartesian coordinate system consists of the real number line plotted on the x-axis and the y axis. The y axis is perpendicular the the x-axis. The two lines intersect at x,ycoordinates at point 0,0

1. natural numbers

2. whole numbers

3. Integers

4. Real numbers

5. Rational numbers Factoring

1. natural numbers Counting numbers, starts at +1 and increases to the right.

2. whole numbers, Add a Zero to the left of the +1 of natural numbers

3. Integers, Add -1 to the left of the zero and continue left to as far as needed.

4. Real numbers, Numbers that have a decimal point in them. Numbers to the right of the decimal point are less than 1,and numbers to the right of the decimal point are 1 or greater than 1.

5. Rational numbers, Numbers that can be expressed as a fraction

Natural Numbers,Whole Numbers and Real Numbers

10/12/17

Definition { } means a set of numbers. The ... is called an ellipsis and means "and so on."

In broader terms the limit of "and so on."
Means go to infinity to the right or - infinity to the left.

For practical purposes infinity is as large (or small) a number as you need to solve the problem you are working on.
In summary infinity cannot precisely defined but is never the less very useful

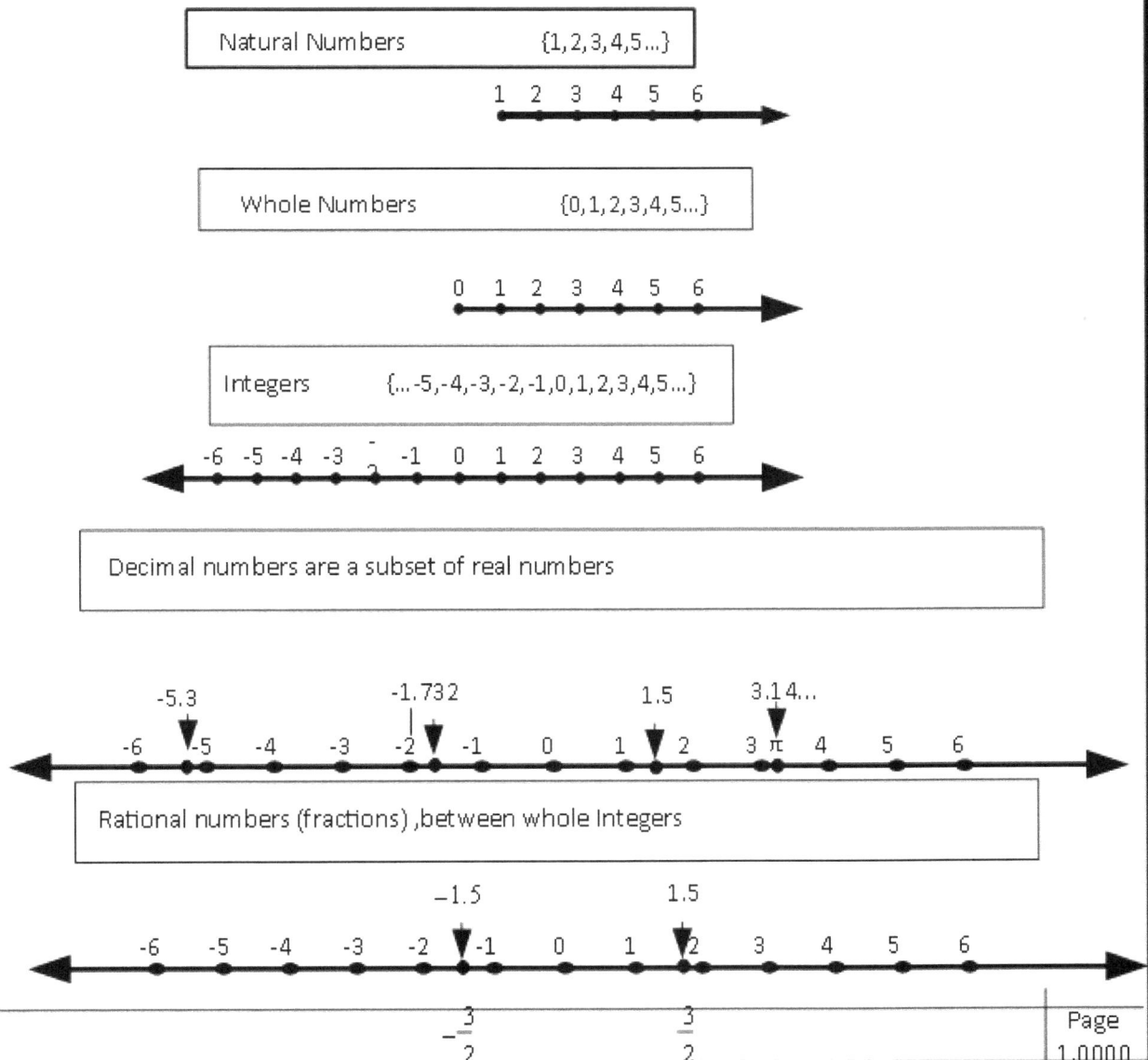

Natural Numbers	$\{1,2,3,4,5...\}$

1   2   3   4   5   6

Whole Numbers	$\{0,1,2,3,4,5...\}$

0   1   2   3   4   5   6

Integers	$\{...-5,-4,-3,-2,-1,0,1,2,3,4,5...\}$

-6 -5 -4 -3 -2 -1 0 1 2 3 4 5 6

Decimal numbers are a subset of real numbers

-5.3   -1.732   1.5   3.14...

-6 -5 -4 -3 -2 -1 0 1 2 3 π 4 5 6

Rational numbers (fractions) ,between whole Integers

−1.5   1.5

-6 -5 -4 -3 -2 -1 0 1 2 3 4 5 6

$-\dfrac{3}{2}$ $\dfrac{3}{2}$

Page
1.0000

Figure 1.1: Real number Types

Properties of Real Numbers	General Algebraic Formula	Real Numbers Examples
Distributive Property Multiplication	$a * (b + c) = (ab + ac)$ $(b + c) * a = (ba + ca)$	$12 * (6 + 5) =$ $(12 * 6 + 12 * 5)$ $= 132$ $(6 + 5) * 12 =$ $(6 * 12 + 5 * 12)$ $= 132$
Identity Property Multiplication	a * 1 = a 1 * a = a	20 * 1=20 = 20 1* 20 = 20
Inverse Property of addition	a + (-a) = 0	4 + (-4)=0
Inverse Property of Reciprocals	a * 1/a = 1	12 * 1/12 = 1

1.1   Rules for factoring-2

Properties of Real Numbers	General Algebraic Formula	Real Numbers Examples
Closure Property	a + b = c unique number	12 + 6 =18
Closure Property	a * b = d unique number	12 * 6 =72
Communicative Property of addition	a + b = b + a	4 + x = x + 4
Communicative Property of Multiplication	ab = ba	12 * 6 = 6 * 12
Associative Property of Addition	(a + b) +c = a + (b + c)	(12 + 6) + 5 = 12 + (6 + 5)
Associative Property of Multiplication	(ab)c = a(bc)	(3*7)5 =3(7*5)

The basic mode of algebra is to

develop rules for handling

unknowns, coefficients

and numbers and exponents.

1.2   Letter Variables

The basic letter variables of Algebra are usually (x, y, and z)

and possibly (a, b, and c) Any letter can be used depending

on the application required.

1.3   State of variables during an Algebra Problem

The Algebraic variable is constant during the duration of the

solution.

Note: In calculus the variables are (v, t, d, and a) may be
     constantly

changing during the problem.

For instance, a drag race car if accelerating at a constant rate

the velocity is continually changing while the acceleration

may be constant in some cases. The time and distance are

changing as well.

1.4 General Characteristics of Algebra

Algebra is a Math that can

transform a formula from a word problem

which is composed of add

terms and subtracted terms into factors of multiples.

When in the multiplied factors format,

we can solve the quadratic equation by

setting the equation to either:

- y (for the graph) or
- 0 (for the x intercepts)

$96 = x(x-4)$	Original Equation
$96 = x^2 - 4x$	Multiply through
$x^2 - 4x - 96 = 0$	Solve for zero
$(x-12)(x+8) = 0$	Factored equation
$(x-12) = 0$	Set factor to zero
$x = 12$	solved
$(x+8) = 0$	Set factor to zero
$x = -8$	solved

1.5   Solving an equation by factoring

Example: Calculate the time the ball takes to hit the ground when thrown upwards from a certain height.

1.6   Zero-Factor property, rule

We can use the zero-factor rule.

When there two factors equal to zero then either factor or both are equal to zero.

Therefore, set each factor equal to zero and solve.

By the Zero factor property

set each factor equal to zero

and solve

$$(x + 1) = 0; \quad x = -1$$
$$(x - 7) = 0; \quad x = 7$$

1.7   Variables in a problem using the same formula

In a further example the constants representing the

variable will change while

the formula will be reused for different

initial and final conditions.

In other words, the formula can be

re-used with different constants but using the same

variables without changing the equation.

In summary in algebra the attribute is

that the variable stays the same for the

duration of the problem to be solved.

Chapter 2

Graphs

2.1 Functionality of graphs in the problem-solving process

In Mathematics graphs are extremely useful to see how a curve behaves. Points such as $(x - \text{intersect})$ and $(y - \text{intersect})$ can usually be seen (given observing that section of the curve) also, as well as the shape of the curve. In algebra a curve can be a straight line, Linear, or a Quadratic,or an Exponential curve resulting in a curved line. The shape of the curve can be a Straight line, A parabola, a Sine Wave, a Tangent wave, etc. In electronics the curve exhibited by a diode, or a transistor, for instance, has a distinctive characteristic curve which is used in designing a circuit.

2.2   Plotting Equations

The Object of plotting equations is to observe the behavior of
the equations. The solution can be often observed, by graphing the equation and observing intersect points and slopes. In
engineering the shape of the equation is useful.

Chapter 3

Plotting two coordinate points

3.1 Cartesian Coordinate System

The Cartesian coordinate system consists of the real number

line plotted on the x-axis and the y axis.

The y axis is perpendicular the x-axis.

The two lines intersect at x, y coordinates at point 0,0

The diagram in the appendix is an example of

the types of real number lines

in the real number system.

Note: The real number

system consists of all the numbers

on the real number line.

In a graph of the real number line

this would represent the x-axis.

The Cartesian coordinate system consists of the real number

line plotted on the x-axis

and the y axis. The y axis is

perpendicular the x-axis.

3.2 Calculate the slope

Slope Problem: Plot (1.-2) and (-4,7) on the Cartesian coordinate system graph.

Plot the x coordinate first then the y-coordinate. (x,y)

The x-coordinate is first in the ordered pair of points. (x,y)

Connect the two points and plot the slope.

- $m = \text{slope} = \frac{\Delta y}{\Delta x}$
- $m = \frac{9}{-5} = -1.8$

Where delta x=change in x and delta y= change in y.

3.3   Graph of plotting two points

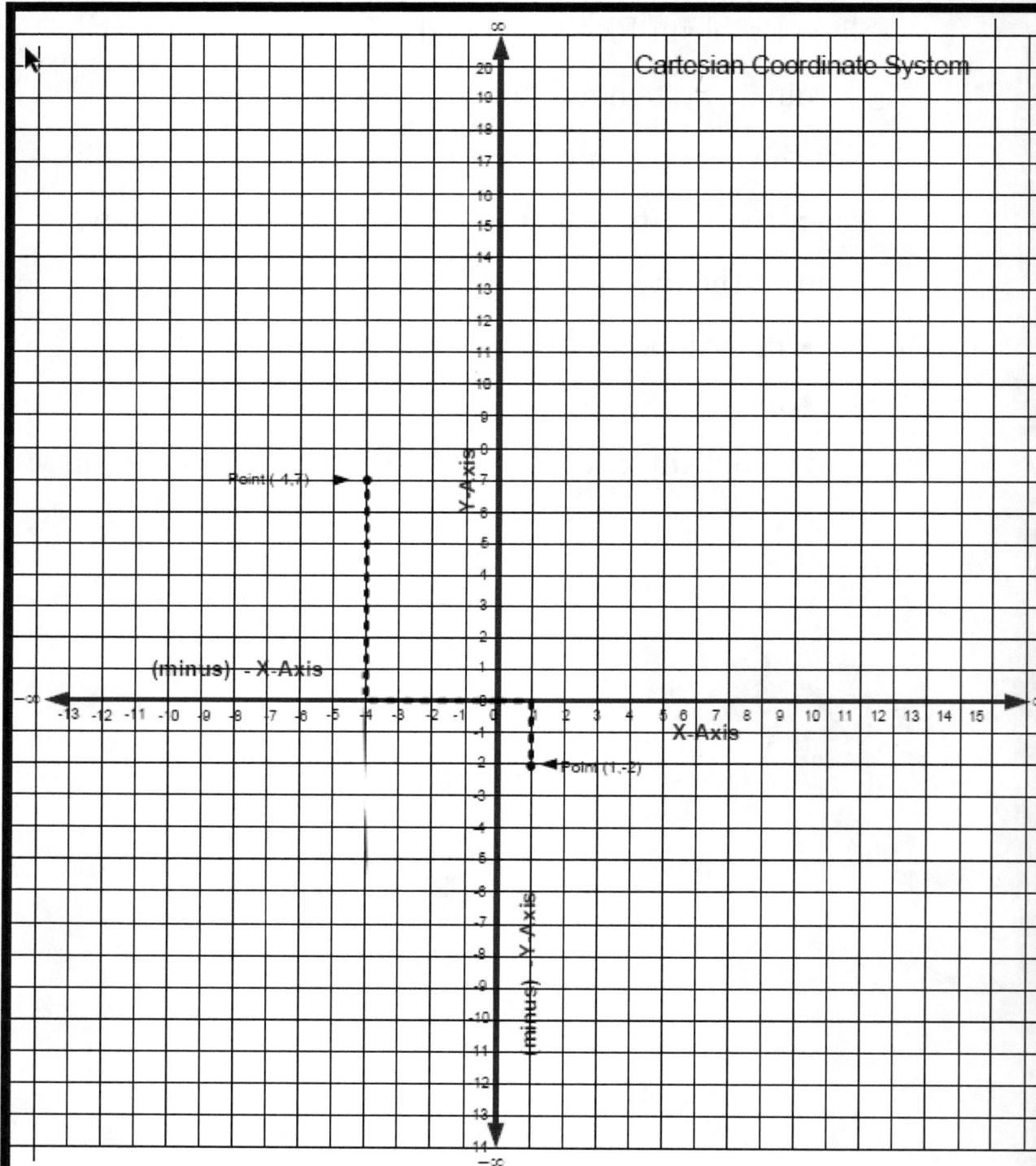

Problem: Plot (1.-2) and (-4,7) on the cartestian coordinate system graph. Plot the x coordinate first then the y-coordinate. (x,y),where the x-coordinate is first in the ordered pair of points.

3.4 Graph of plotting connecting the two points

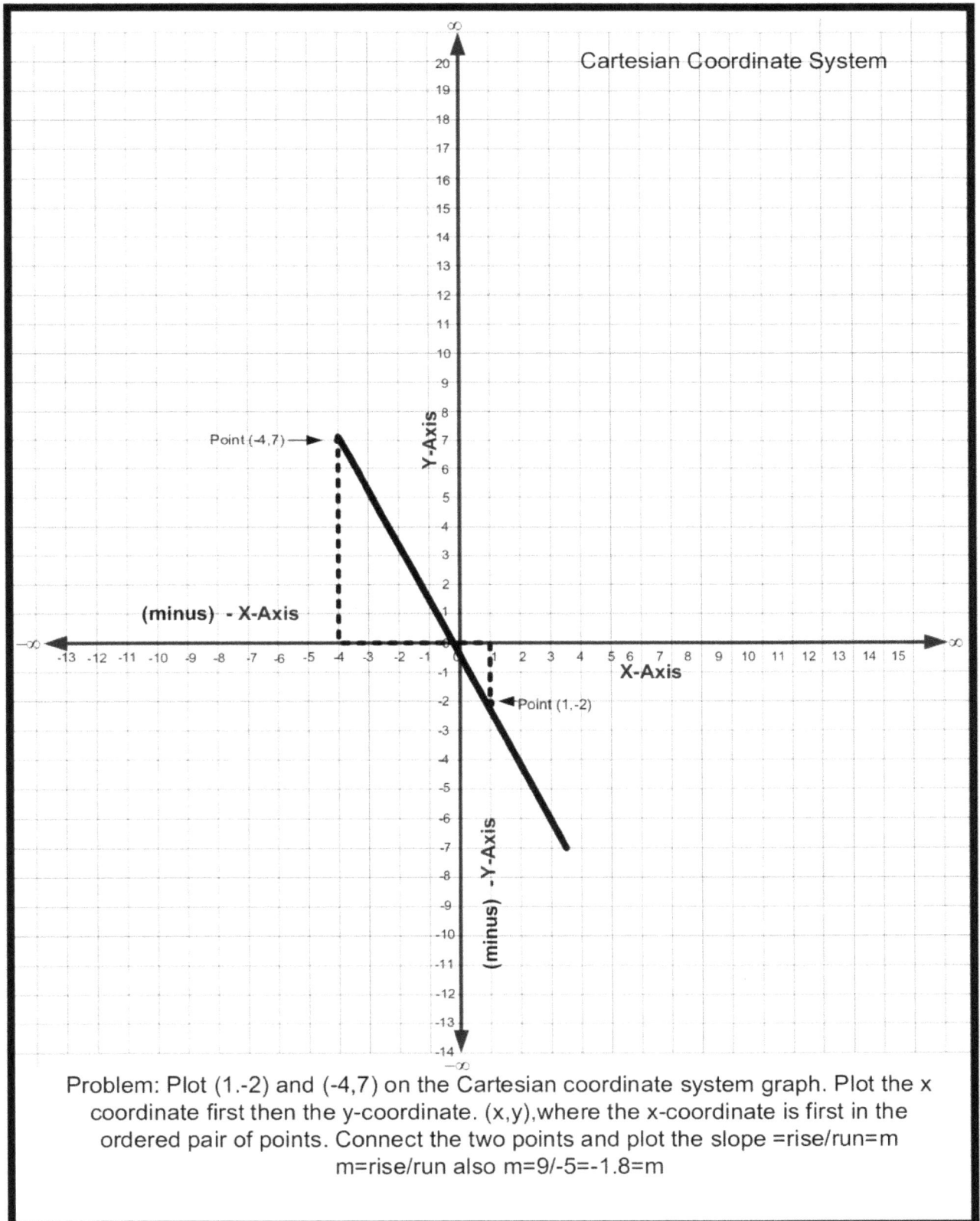

Cartesian Coordinate System

Point (-4,7)

Y-Axis

(minus) - X-Axis

X-Axis

Point (1,-2)

(minus) -Y-Axis

Problem: Plot (1.-2) and (-4,7) on the Cartesian coordinate system graph. Plot the x coordinate first then the y-coordinate. (x,y),where the x-coordinate is first in the ordered pair of points. Connect the two points and plot the slope =rise/run=m
m=rise/run also m=9/-5=-1.8=m

Chapter 4

Polynomials Basic Forms

4.1 Polynomials

Polynomials are basically in 3 separate formats

Which use a combination of numbers and letters

raised to some power.

Monomial $3x^2$

Binomial $(x + 5)$

Trinomial $x^2 + 4x + 4$

Polynomial $x^4 + 2x^3 + 3x^2 + 5x + 7$

$x^4 + 2x^3 + 3x^2 + 5x + 7$

Has 5 terms greater than 3 terms and is simply named a polynomial.

In general all the expressions list above are also call polynomials.

4.2   Monomials

A monomial ,mono meaning one, consists of a letter representing a variable

$$x, y, z, a, v, t, a, b, c$$

and a number preceding the variable letter variable does not show, and the one is not necessary, then the number preceding it is assumed to be one. In general, the 1 exponent, is not specified but assumed, therefore, the one is not necessary to express the x. Therefore $x = 1x$

Examples of monomials are:

$$x, \ 3x^2, \ 10y$$

4.3   Binomials

Binomials (Bi meaning two) are expressions of two monomials separated by a plus (+) or minus (-) operators, and enclosed by parenthesis.

4.4   Examples of binomials

$(x - 3)$ is separated by a minus sign and is one entire unit designated as a binomial. $(3x+5)$ is another example of a binomial. Note the parenthesis around a binomial. In general, a binomial is $(ax \pm b)$

4.5   Trinomials

Trinomials (tri for three) means 3 terms. The usual trinomial form we see in algebra is of the form $ax^2 + bx + c$ it may in many cases be factored into two binomial factors i.e. $(x \pm a)(x \pm b)$ where x is the variable and a and b are constants. A typical example of a trinomial is $x^2 + 4x + 4$ where a=1, b=4, and c=4. Much of learning algebra in the early part of study is learning the rules and methods of factoring.

4.6   Trinomials continued

More advanced applications come later in the learning and involve understanding and applications of algebra.

Note that algebra is used in Physics, Chemistry, Trigonometry and Calculus.

The algebraic rules do not change. The main difference is that functions will be applied along with variables, such as sin, cos, tan, csc, sec and cot, in trigonometry.

In calculus the variables are changing during the solution to the problem, whereas the variable does not change during the solution in Algebra.

Also the variables in Physics v' velocity a, acceleration t'time and d,distance and have specific physical properties. as in d=rt distance = rate * time

4.7   Polynomials with an exponent of 4 or higher

For example

$$x^4 + 3x^2 + 2x + 15$$

This Polynomial may be factorable. In the case it is not factorable it is said to be in its prime state.

4.8 Examples

The basic mode of algebra is to develop rules for handling unknowns coefficients and numbers and exponents.

The advantage in learning algebra is that the same algebraic rules apply to trigonometry and calculus.

Trig and calculus layer on more attributes on algebra. and use the basic rules of Algebra.

4.9 Letter Variables

The basic letter variables of Algebra are usually $x.y.z$ and a, b, c Any letter can be used depending on the application required

4.10   Factor out any monomial term

Factor out any common monomial term

Then solve the remaining trinomial.

For example, $3x^2 + 12x + 12$. The common factor is 3.

Factoring gives us $3(x^2 + 4x + 4)$.

This can be factored to $3(x + 2)(x + 2)$ or $3(x + 2)^2$

using factoring techniques.

Chapter 5

Factor by Grouping

5.1   Summary for Factor by Grouping

When solve quadratic equations with the
coefficient of the first term,
or the squared term we:
Factor out any monomial term common
to each term in the quadratic equation.

1. In the case of

$$-24x^2 - 42x + 45 \text{ we can factor out } -3$$

$$\text{giving us} -3(8x^2 + 14x - 15)$$

2. Note that we factored a -3 instead of a $+3$ giving us
   a positive coefficient for the first term.

3.   If the coefficient of the squared term still does
   not equal one then factor by grouping.

4. We need to factor $(8x^2 + 14x - 15)$. Since the coefficient

5.2   Steps for factor by grouping

1. Solve

$$-24x^2 - 42x + 45 \text{ we can factor out } -3$$
$$\text{giving us} - 3(8x^2 + 14x - 15)$$

2. Multiply the coefficient of the first term times the third term integer,

3. Split the middle term coefficient $+14$ into two terms that are equivalent to the middle term

4. Group into 2 binomial terms

5. Factor any common binomial

6.   3x is the largest factor that can be factored out of all 3 terms. Note: We divide out -3x instead of $+3x$ because we need to have the first term inside the parenthesis be a positive value.

7. Factored out binomial -3

$$-3(8x^2 + 14x - 15)$$

5.3   Split the middle term into 2 equivalent terms

1. In a quadratic equation we multiply two terms to get the third term.

2. Factoring is the opposite of multiplying so the third term cam be split into factors.

3. Multiply the first term coefficient time the third term integer,

4. Multiplying

$$8 \bullet (-15) = -120$$

5. In a quadratic equation we add two terms to get the coefficient of the middle term.

6. The factors of the third term need to add up to the middle term.

5.4 Factor −120

1. The easiest way to factor is to divide the number by a small prime number, such as 2, 3, or 5,sometimes 7 or a higher prime number.

2. nm = -120 If we start by dividing by 2 and continue dividing by 2 we get

$$-1 \bullet 2 \bullet 2 \bullet 2 \bullet 15$$

3. Since we have run out of $2's$ we divide by the next prime number 3

$$-1 \bullet 2 \bullet 2 \bullet 2 \bullet 3 \bullet 5$$

5.5   Combine the two terms that add up to $+14$

1. factors
$$-1 \bullet 2 \bullet 2 \bullet 2 \bullet 3 \bullet 5$$

2. Test
$$8 + (-15) = -7 \quad \text{Fails}$$

3. Test
$$-6 + 20 = 14 \quad \text{Passes}$$

4. Since

$$-6 + 20 = 14 \quad \text{Middle term and } (-6 \bullet 20) = -120 \text{ Third term}$$

5. This is the correct combination of factors

5.6   Rewrite the middle term 14 to $-6{+}20$ equivalent term

1.   Now we have 2 factors that add up the middle term we can rewrite the equation:

2. (Don't forget to save the -3x factor for the final list of factors)
$$-3(8x^2 + 14x - 15) \ = -3(8x^2 + 20x - 6x - 15)$$

3. Rewrite the middle term
$$-3(8x^2 - 6x + 20x - 15)$$

4. Group two sets of terms into factors
$$-3(8x^2 - 6x)(+20x - 15)$$

5. Factor out monomials.
$$-3 \quad ((2x)(4x - 3) + ((5)(4x - 3))$$

6. Factor out binomial $(4x - 3)$
$$-3 \bullet ((4x - 3) \bullet (2x + 5))$$

5.7 Check by using FOIL to multiply

1. FOIL multiply

$$-3 \bullet \left(8x^2 + 20x - 6x - 15\right)$$

$$-3 \bullet \left(8x^2 + 14x - 15\right)$$

2. Multiply

$$-3 \bullet \left(8x^2 + 14x - 15\right)$$

3. The calculation checks out.

$$-24x^2 - 42x + 45 \quad = -24x^2 - 42x + 45 \quad \text{Original equation}$$

5.8 Quadratic Equation using logic diagram.

Factor by grouping is a more direct solution to this problem.

$$12x^2 - x - 20$$

Multiple the first term coefficient and the third term constant.

$$12 * (-20) = -240$$

$$\text{Factor}[-240] = (-1)\left(2^4\right)(3)(5) = (-16)(15)$$

Combine the factors that add up to the coefficient of the middle term.

$$-16 \text{ and } 15 \text{ add up to } -1$$

which is the correct coefficient of the middle term.

now we rewrite the middle term resulting in 4 terms.

$$12x^2 - x - 20$$

becomes

$$12x^2 - 16x + 15x - 20$$

Note : $-16x + 15x$ is equivalent to $-x$ when added Now Group the terms and factor.

$$\left(12x^2 - 16x\right) + (15x - 20)$$

$$4x\left(3x - 4\right) + 5\left(3x - 4\right)$$

Now factor out the $(3x - 4)$ resulting in

$$(3x - 4)\left(4x + 5\right)   \text{Answer}$$

Chapter 6

Factoring using the Difference of squares method

6.1 Difference of squares is a Special Factoring product.

$(a + b)(a - b) = a^2 - b^2$

$(a + b)(a - b) = a^2 + ba - ba - b^2$

The middle term drops out when you multiple two opposite binomials.

therefore $a^2 - b^2 = (a + b)(a - b)$

This product is called **The Difference of Squares** product.

Examples of difference of squares

$$x^2 - 9 = (x - 3)(x + 3)$$

$$x^4 - 1 = \left(x^2 - 1\right)\left(x^2 + 1\right)$$

Intermediate step

Note $\left(x^2 - 1\right)$ is also a perfect square

Therefore:

$$x^4 - 1 = (x - 1)(x + 1)(x^2 + 1)$$

Another example of Differences of squares

$$x^2 - 4y^2 = (x + 2y)(x - 2y)$$

6.2   Difference of squares is a Special Factoring product.

Examples of difference of squares

$$x^2 - 9 = (x - 3)(x + 3)$$

$$x^4 - 1 = (x^2 - 1)(x^2 + 1)$$

Intermediate step

Note $(x^2 - 1)$ is also a perfect square

Therefore:

$$x^4 - 1 = (x - 1)(x + 1)(x^2 + 1)$$

Another example of Differences of squares

$$x^2 - 4y^2 = (x + 2y)(x - 2y)$$

Chapter 7

Special product factoring

7.1 Perfect Square Trinomials

The special products rule for an algebraic term ,squared.

$$(a + b)^2 = a^2 + 2ab + b^2$$
$$(a - b)^2 = a^2 - 2ab + b^2$$

7.2 Perfect Square Trinomial

A perfect square trinomial is the square of a binomial

$x^2 + 10x + 25$ is a perfect square trinomial.

25 is a perfect square.

the middle term test: 2 times the product of the square root of the first and third term.

Note: $\sqrt{x^2} = x$ $\sqrt{25} = 5$

$$2 * x * 5x = 10x$$

The middle term is correct.

The middle term test passes

Therefore, the binomial consists of the addition of the two square roots squared.

$(x+5)^2$ *answer*

7.3   Perfect Square Trinomial

$x^2 - 22x + +121$ is a perfect square trinomial.

121 is a perfect square.

the middle term test: 2 times the product of the square root of the first and third term.

The result of a square root $\sqrt{121} = -11$ or 11
   Note : Two solutions

Note: $\sqrt{x^2} = x$ $\sqrt{121} = 11$
Note: $-\sqrt{x^2} = -x$ $\sqrt{121} = -11$

$$2 * x * -11x = -22x$$

The middle term is correct.

The middle term test passes

Therefore, the binomial consists of the subtraction of the two square roots squared.

$(x-11)^2$ *answer*

Chapter 8

Factoring decisions logic flow chart

8.1 Factoring facts to consider

The method to solve this equation intuitively look at several factors:

Is the coefficient of the first term one?

If the coefficient greater than one? Do factor by grouping.

Is the middle term coefficient large? Then one of the m's or n's will be large,

Are the signs of the second term and third positive?

Is the coefficient of the third term negative? The factored terms will have opposite signs Plus-Minus or Minus-Plus

If the third term coefficient is positive both signs of the binomials will be the same. Plus-Plus or Minus-Minus

Chapter 9

Process Graph for solving a Quadratic Equation

1. The purpose of the following diagram is to show in programming diagram format the decisions made to solve this trinomial into binomial factors

2. The rounded rectangle shows the start of the process.

3. The diamond represents a decision to be made in solving the problem.

4. The Rectangle shows the result of the solution.

5. The shaded in shapes show the path of the decision for this example

6. The flow diagram is shown in order to develop a more logical approach to factoring by examining the middle term and the sign of the third term.

Diagrams of factoring rules

Monday, June 29, 20

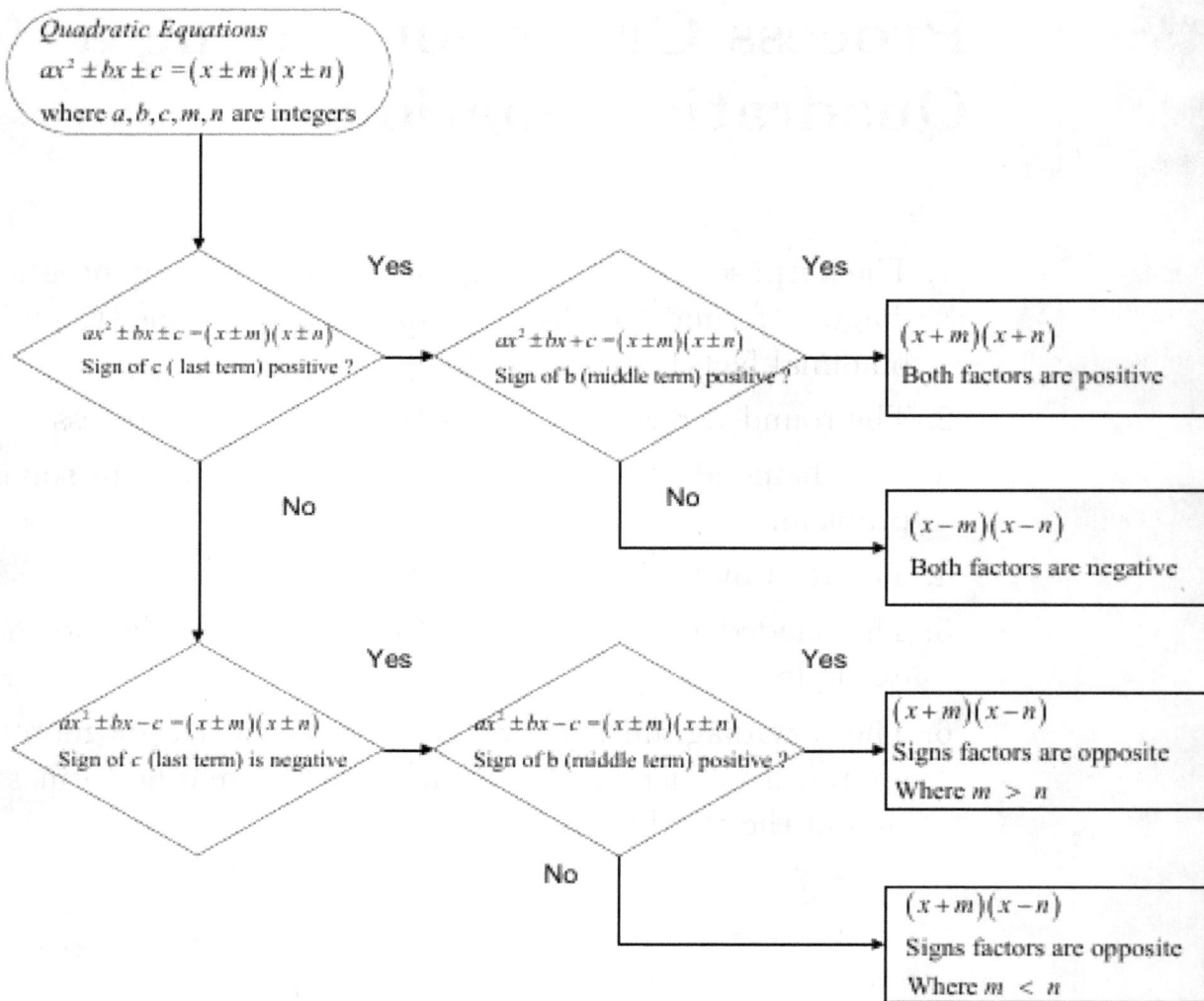

$Quadratic\ Equations$

$ax^2 \pm bx \pm c = (x \pm m)(x \pm n)$

where a, b, c, m, n are integers

$ax^2 \pm bx \pm c = (x \pm m)(x \pm n)$
Sign of c (last term) positive ?

Yes

$ax^2 \pm bx + c = (x \pm m)(x \pm n)$
Sign of b (middle term) positive ?

Yes

$(x + m)(x + n)$
Both factors are positive

No

$(x - m)(x - n)$
Both factors are negative

No

$ax^2 \pm bx - c = (x \pm m)(x \pm n)$
Sign of c (last term) is negative

Yes

$ax^2 \pm bx - c = (x \pm m)(x \pm n)$
Sign of b (middle term) positive ?

Yes

$(x + m)(x - n)$
Signs factors are opposite
Where $m > n$

No

$(x + m)(x - n)$
Signs factors are opposite
Where $m < n$

Page 1

Figure 9.1: Factoring-rules-template

Chapter 10

Small Middle term

$ax^2 \pm bx \pm c = (x \pm n)(x \pm m)$
$12x^2 - x - 20$

 The Coefficient of the middle term is small compared to the coefficients of the first term and the constant in the third term. Therefore, choose small terms in binomials and coefficients will be close in value.

Sign of c last term positive? No

Sign of b middle term positive? No

Sign factors are opposite where $m < n$

See logic chart for highlighted decision path,

Saturday, May 4, 2019

Factoring Rules for small middle term coefficient, compared to the third term.

Quadratic Equations
$ax^2 \pm bx \pm c = (x \pm m)(x \pm n)$
where a, b, c, m, n are integers.

$12x^2 - x - 20 = (ax + m)(bx - n)$

Since the middle term is small use smaller integers to tend to "balance" the pluses and minuses creating a small middle term

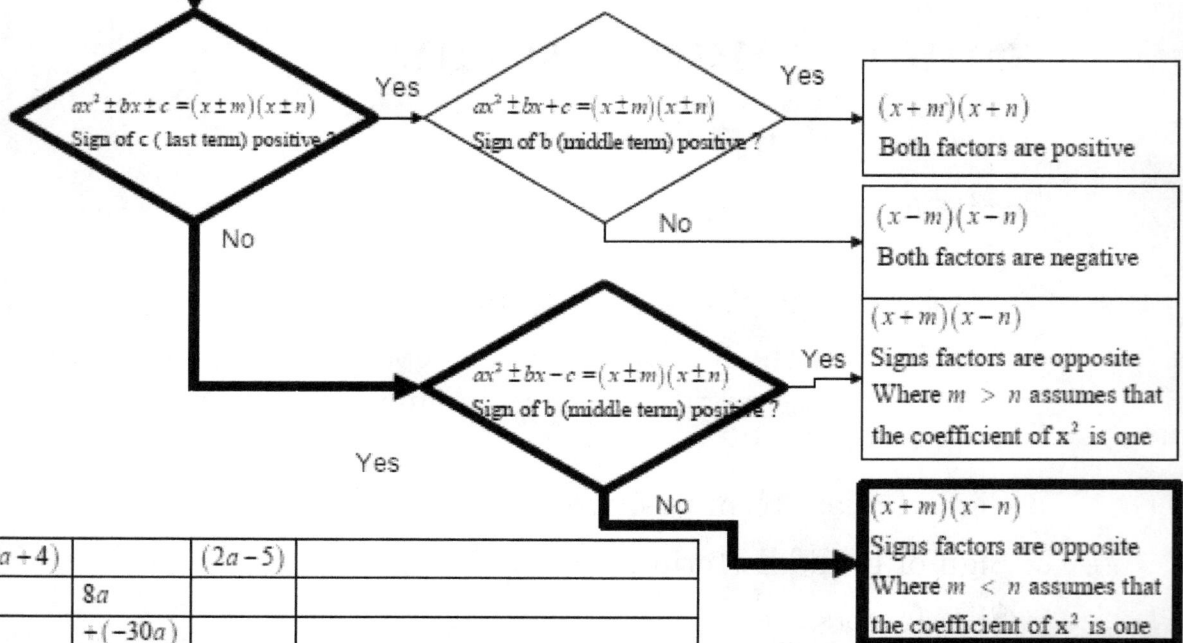

$ax^2 \pm bx \pm c = (x \pm m)(x \pm n)$
Sign of c (last term) positive ?

Yes

$ax^2 \pm bx + c = (x \pm m)(x \pm n)$
Sign of b (middle term) positive ?

Yes

$(x + m)(x + n)$
Both factors are positive

No

$(x - m)(x - n)$
Both factors are negative

No

$ax^2 \pm bx - c = (x \pm m)(x \pm n)$
Sign of b (middle term) positive ?

Yes

$(x + m)(x - n)$
Signs factors are opposite
Where $m > n$ assumes that the coefficient of x^2 is one

Yes

No

$(x + m)(x - n)$
Signs factors are opposite
Where $m < n$ assumes that the coefficient of x^2 is one

$(6a + 4)$		$(2a - 5)$					
	$8a$						
	$+(-30a)$						
	$= -22a$		incorrect middle term	Factor 12	12	1	Very large term
					6	2	Try first time
$(3a + 4)$		$(4a - 5)$			4	3	Try Second time
	$16a$						
	$-15a$			Factor -20	-20	1	Very large term
	$= (a)$		wrong sign, so reverse the sign only		20	-1	Very large term
$(3a - 4)$		$(4a + 5)$			10	-2	Large term
	$-16a$				-10	2	Large term
	$+15a$				4	-5	Try first time
	$= (-a)$		correct middle term		-4	5	Reverse the signs

$12x^2 - x - 20 = (4x + 5)(3a - 4)$ answer

Page 1

Figure 10.1: Factoring Logic Chart

Chapter 11

Linear equations

Three solution types including the graphing of the equations.

11.1 Solutions a system of linear equations

- In order to better understand the characteristics of linear equations , we can solve them as
- Solving by addition, Check by Graphing the two equations.
- Solving by Cramer's rule.
- Check by Graphing the two equations.

11.2   Simple-linear-equations

11.2.1   Linear equations 1 variable x.

The following pages show solutions for linear equations.

Linear are solved for the variable x.

Since y=f(x)

We substitute the value of x into the original

equation and solve for the value of y giving us the

coordinates of an intersection

given 2 linear equations.

11.3   Linear Equations one variable x

Solve:

$$2x + 3 = 17$$

$$2x + 3 - 3 = 17 - 3$$

add -3 to both sides to get to isolate the $2x$ Use the additive inverse -3

$$2x = 14$$

$$\frac{2x}{2} = \frac{14}{2}$$

Divide both sides by 2 to isolate the x Use the multiplicative inverse $\frac{1}{2}$

$$x = 7 \quad \text{Solution.}$$

11.4   Substitute in the original equation

$$2x + 3 = 17$$

$$2(7) + 3 = 17$$

$$17 = 17$$

The solution (17) ; checks out

11.5 Solving Linear Equations Example 2

Solving Linear Equations 1 variable x,

$$2x - 17 + x = 3x + 1 + 2x$$

$$3x - 17 = 5x + 1$$

Solve each side then isolate the x by combining both sides.

$$3x - 17 - 5x = 5x + 1 - 5x$$

$$-2x - 17 = 1$$

$$-2x - 17 + 17 = 1 + 17$$

$$-2x = 18$$

$$\frac{-2x}{2} = \frac{18}{2}$$

$$-x = 9$$

$$x = -9$$

Solution $x = -9$ Check by substituting in the original equation.

$$2x - 17 + x = 3x + 1 + 2x$$

$$2(-9) - 17 + (-9) = 3(-9) + 1 + 2(-9)$$

$$-18 - 17 - 9 = -27 + 1 - 18$$

$$-44 = -44$$

it checks out !!

The intersection was calculated at

$$\{x, y\} = \{-9, -44\}$$

To find the y-intersection set x=0 and calculate:

$$\{x, y\} = \{0, 1\}$$

and

$$\{x, y\} = \{0, -17\}$$

Chapter 12

Addition method with the result 8,-3

12.1 Linear-system-Solve-by-Addition

Solve the following linear system.

$$3x + 11y = -9$$
$$7x + 6y = 38$$

$$\left| \begin{array}{l} 3x + 11y = -9 \\ 7x + 6y = 38 \end{array} \right|$$

Multiply the first equation by 7
Multiply the second equation by -3
The $x's$ drop out when you add the two equations.

$$(3x + 11y = -9) \bullet 7$$
$$(7x + 6y = 38) \bullet -3$$

$$(3x + 11y = -9) \bullet 7 = (21x + 77y = -63)$$
$$(7x + 6y = 38) \bullet -3 = (-21x - 18y = -144)$$

Add the two equations

$$21x + 77y = -63$$
$$-21x - 18y = -144$$

$$-\;-\;-\;-\;-\;-\;-\;-\;-\;--$$

$$-59y = 177$$
$$y = -3$$

Substitute the value of y in either equation and solve for x.

$$3x + 11\,(-3) = -93x - 33 = -9$$
$$3x = 24$$
$$x = 8$$

The solution is (x, y)=(8,-3)

Check

$3x + 11y = -9$

$3\,(8) + 11\,(-3) = -9$

$24 - 33 = -9$

$-9 = -9$ It checks !!

12.2 Graphical Solution

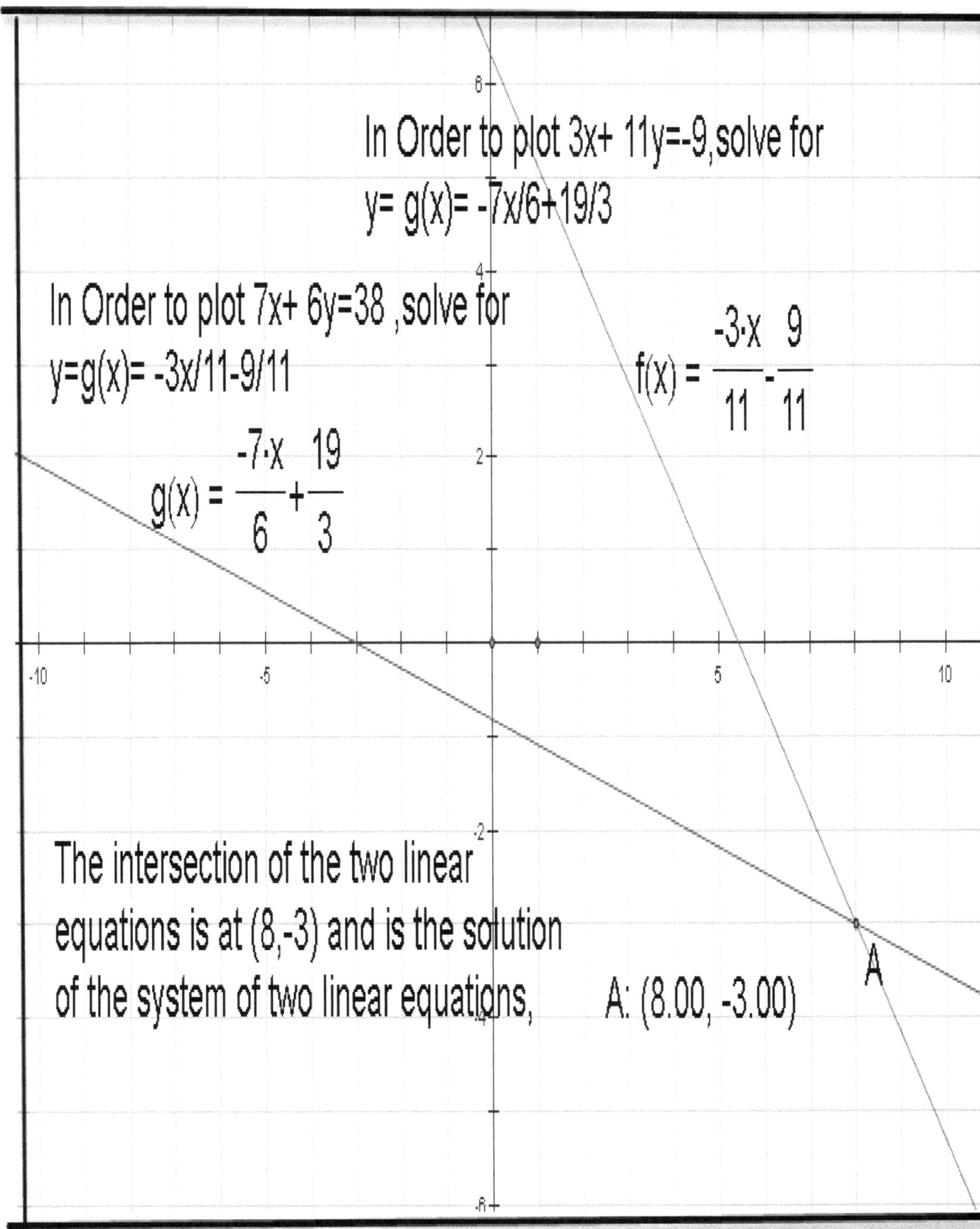

In Order to plot 3x+ 11y=-9,solve for

y= g(x)= -7x/6+19/3

In Order to plot 7x+ 6y=38 ,solve for

y=g(x)= -3x/11-9/11

$$f(x) = \frac{-3 \cdot x}{11} - \frac{9}{11}$$

$$g(x) = \frac{-7 \cdot x}{6} + \frac{19}{3}$$

The intersection of the two linear equations is at (8,-3) and is the solution of the system of two linear equations,   A: (8.00, -3.00)

A

12.3   Linear equations

Solutions for linear equations.

$$4(2x + 1) - 29 = 3(2x - 5)$$

In this case the two equations Intersect since their slopes are not equal.

$$y_1 = 4(2x + 1) - 29$$

$$y_2 = 3(2x - 5)$$

12.4   Slopes of left and right side of the equation.

The lines intersect at

$$y_1 = y_2$$

Left side $y_1 = mx + b$ where m is the slope

$$4(2x + 1) - 29$$

reduces to

$$8x + 4 - 29 = 8x - 25$$

where the slope is 8

Right side $y_2 = mx + b$ where m is the slope

$$3(2x - 5)$$

reduces to

$$6x - 15$$

where the slope is 6 Therefore the 2 lines intersect, since the slopes would have to be equal in the case of parallel
lines or overlapping lines. If the slopes are not
equal the lines would have to intersect at some point,

12.5 Identify two other line types

Note: if the slopes were equal, they would be parallel or overlap.

The lines would be parallel if slope is the same and the $b = y - intercept$ were different. Note if the y-intercepts were the same and the slopes were equal then the lines would overlap.

12.6 Linear equation set example

Linear equation set example, solving for x, y, and y-intercepts and graphing the solutions

$$4(2x + 1) - 29 = 3(2x - 5)$$

Distribute the coefficient over the binomial in the parenthesis to clear the parenthesis from the equation.

$$8x + 4 - 29 = 6x - 15$$

$$8x - 25 = 6x - 15$$

$$8x - 25 - 6x = 6x - 6x - 15$$

Collect all the x terms on the left side of the equation and add the terms.

$$2x - 25 = -15$$

Collect all the *digits* terms on the right side of the equation and add the terms.

$$2x - 25 + 25 = -15 + 25$$

$$2x = 10$$

$$x = 5$$

Solution $x = 5$ Next we solve for y by substituting the value of $x = 5$ in either side of the original equations. Note that $y = f(x)$ means that the value of y

depends on the value of x. We have the value of y from the solution for

the equation therefore:

$$4(2x + 1) - 29 = 3(2x - 5)$$

Solving the left side of the equation:

$$y = 4(2(5) + 1) - 29 = 44 - 29 = 15; y = 15$$

Solving the right side of the equation:

$$y = 3(2x - 5)$$

$$y = 3(10 - 5) = 15; y = 15$$

This make sense since at

$$(x, y) = (5, 15)$$

The equations are equal Since the lines intersect this

$$(5, 15)$$

is a unique solution.

12.7 Graphing solution.

We demonstrate this by graphing the lines
represented by each side of the equation
and observing that the intersection is
at

$$(x, y) = coordinate\ points\ (5, 15)$$

12.8 Plot y-intercepts.

Another calculation we make is plotting the y-intersect. By definition the y-intersect is at x=0. Therefore, we solve the left side of the equation by substituting 0 for any x term.

$$y = 4(2x + 1) - 29$$

$$y = 4(0 + 1) - 29 => y = -25$$

for the left side of the equation

also, substitute 0 for any x term on the right side of the equation

$$y = 3(2x - 5)$$

$$y = 3(0 - 5) => y = -15$$

The y intersecting points are

$$(x, y) = (0, -25)$$

for one side of the equation and

$$(x, y) = (0, -15)$$

for the other side of the equation.

1

Intersect Point
for the two
equations

The intersect point at {5, 15}
is the solution to the system
of equations.
4(2x+1)-29=3(2x-5)

$$y_1 = 4(2x+1) - 29$$

$$y_2 = 3(2x-5)$$

g(x) = 3·(2·x-5)   f(x) = 4·(2·x+1)-29

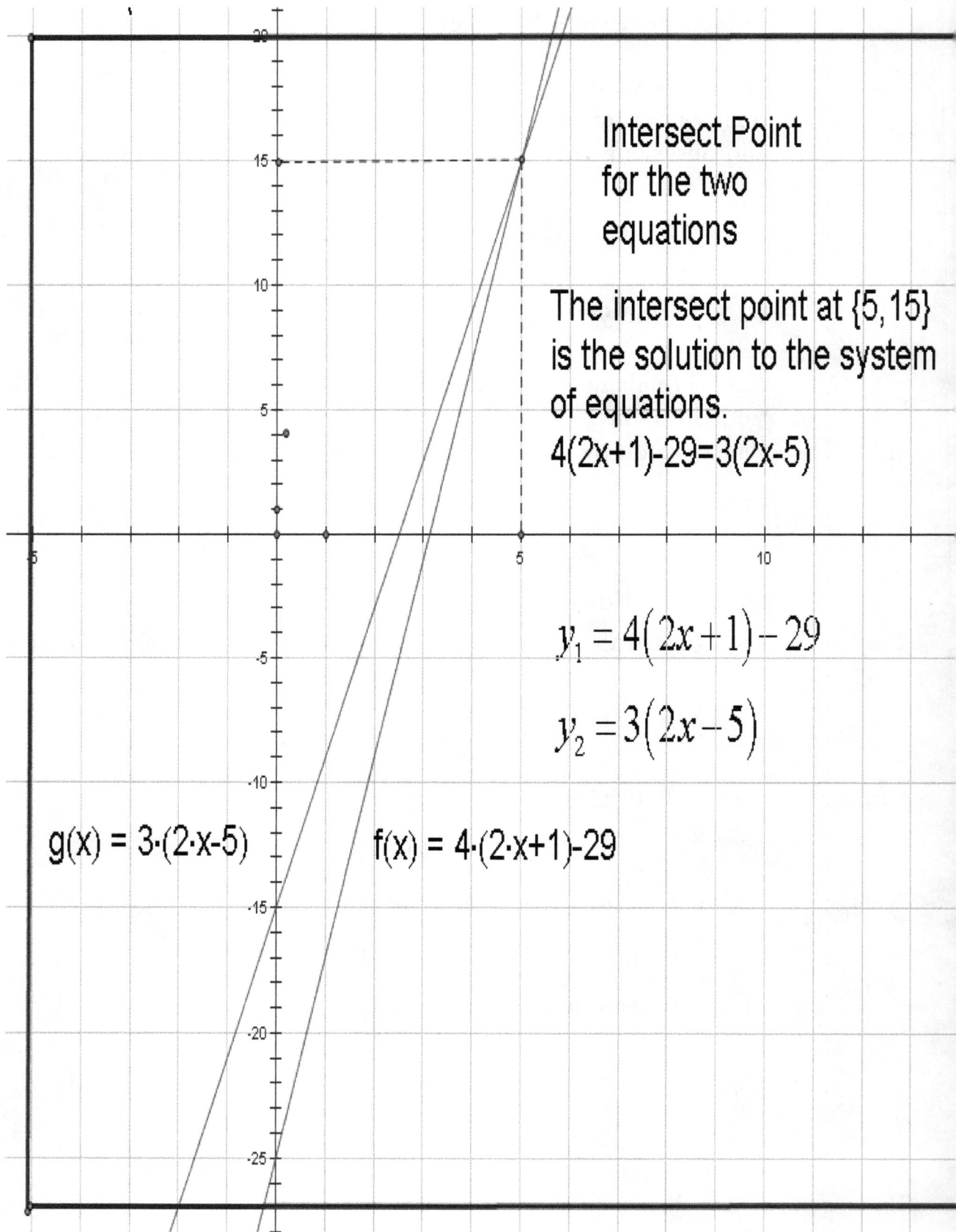

Figure 12.2: Intersect of 2 linear equations

12.9 Another-linear-equation

Linear-quations

Solving Linear Equations Example

$$2x - 17 + x = 3x + 1 + 2x$$

$$3x - 17 = 5x + 1$$

Solve each side separately

$$3x - 17 - 5x = 5x + 1 - 5x$$

$$-2x - 17 = 1$$

$$-2x - 17 + 17 = 1 + 17$$

$$-2x = 18$$

$$\frac{-2x}{2} = \frac{18}{2}$$

$$-x = 9$$

$$x = -9$$

Solution $x = -9$ Solve for y_1

$$y_1 = 2x - 17 + x$$

$$y_1 = 2(-9) - 17 + (-9)$$

$$y_1 = -44$$

Solve for y_2

$$y_2 = 3x + 1 + 2x$$

$$y_2 = 3(-9) + 1 + 2(-9)$$

$$y_2 = -44$$

$$2(-9) - 17 + (-9) = 3(-9) + 1 + 2(-9)$$

Solve for Both sides of the equation

Check by substituting in the original equation.

$$2x - 17 + x = 3x + 1 + 2x$$

$$2(-9) - 17 + (-9) = 3(-9) + 1 + 2(-9)$$

$$-18 - 17 - 9 = -27 + 1 - 18$$

$$-44 = -44$$

it checks out !! Also substituting the value for x (-9) yields (-44) in both sided of the equation then $y = (-44)$ and the intersecting point is $\{-9, -44\}$ The intersection was calculated at

$$\{x, y\} = \{-9, -44\}$$

To find the y-intersection set x=0 and calculate:

$$\{x, y\} = \{0, 1\}$$

and

$$\{x, y\} = \{0, -17\}$$

The x-intersect point of equations E,D is calculated by setting each equation to zero. ←

E: (-0.20, 0.00)

D: (5.66, 0.00)

E

D

$q(x) = 3 \cdot x + 1 + 2 \cdot x$

$h(x) = (2 \cdot x - 17) + x$

The intersect point at {-9,-44} is the solution to the system of equations: (2x-17)+x and 3x+1+2x

A: (-9.00, -44.00)

A

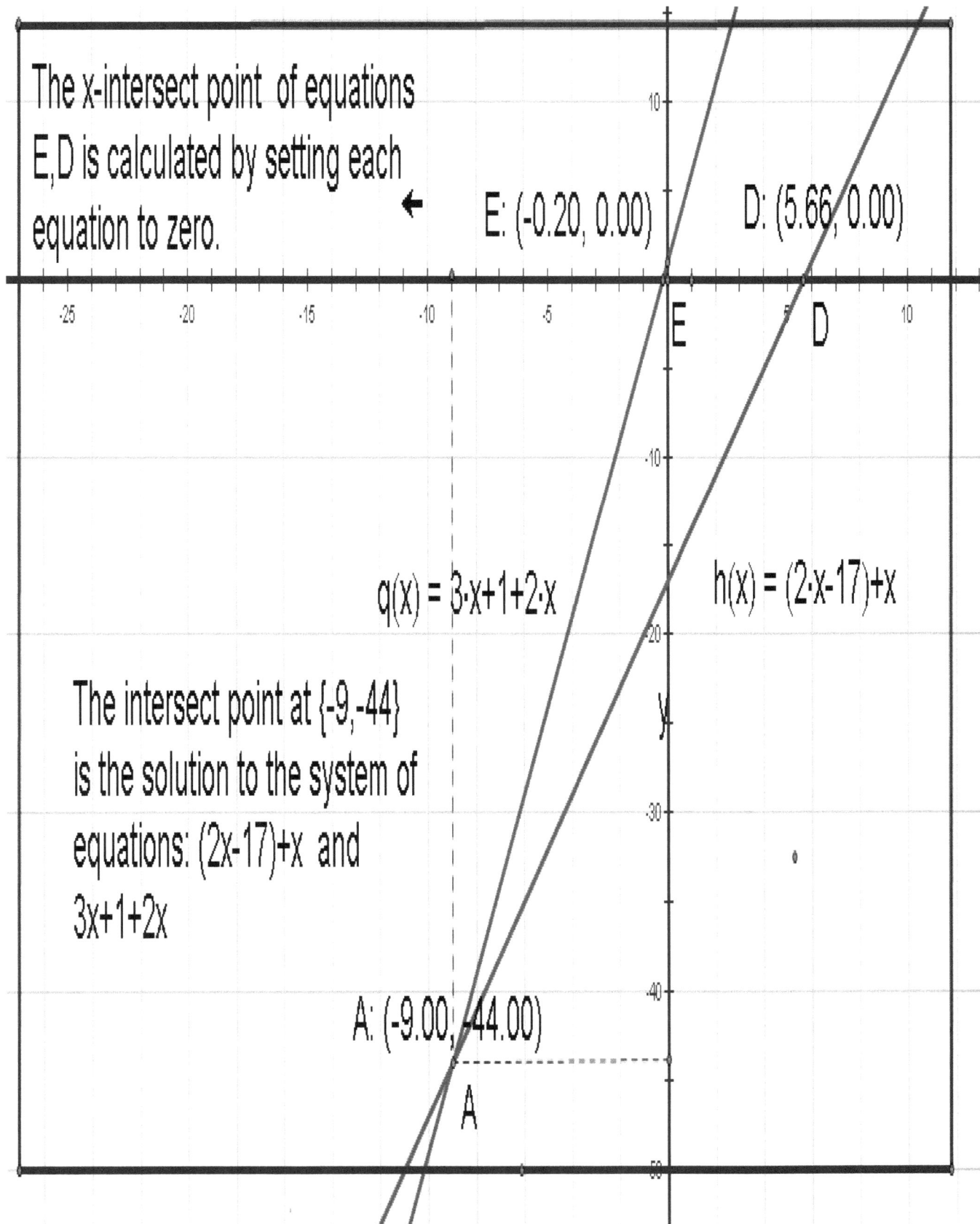

Figure 12.3: Another example of system of 2 equations

12.10   Solutions for two Linear equations

- Intersect,Parallel,and Overlap resulting in
- Consistant Inconsistant and Dependent

When we have two linear equations the solutions can be one of 3 solutions.

- Intersect
- Parallel
- Overlap

A linear equations system has the format: $y = mx + b$

Where the value of y is the dependent variable on x for its value.

x is the independent variable.

b is the y intercept at $x = 0$ creating a point $0, b$

m is the slope and equals

$$\frac{rise}{run} = \frac{\Delta y}{\Delta x} = \frac{vertical\ change\ in\ y}{horizontal\ change\ in\ x}$$

Intersect $m_1 = m_2$ at one point ; only; $b_1 \neq b_2$ Parallel $m_1 = m_2$ at every point $b_1 = b_2$ different $y - intercept$ Overlap $m_1 = m_2$ at every point; $b_1 = b_2$ same $y - intercept$

12.11 Three-types-Linear

Type of Equations	Equation	x, y or Remainder	Number of Solutions
Consistent (Intersecting)	$y = -x - 1$ $4x - 3y = 24$	$\{3, -4\}$	One solution
Inconsistent (Parallel)	$3y - 2y = 6$ $6x - 4y = 24$	x dropped out y dropped out Remainder 0=6 False	No Solution
Dependent (Overlapping)	$y = 3x - 2$ $15x - 5y = 10$	x dropped out y dropped out Remainder 0=0 True	

Consistent

lines intersect at one point Inconsistent lines are paralell and do not intersect Inconsistent lines have the same slope Dependent lines overlap intersect all all points Dependent lines have an infinite number of solutioins

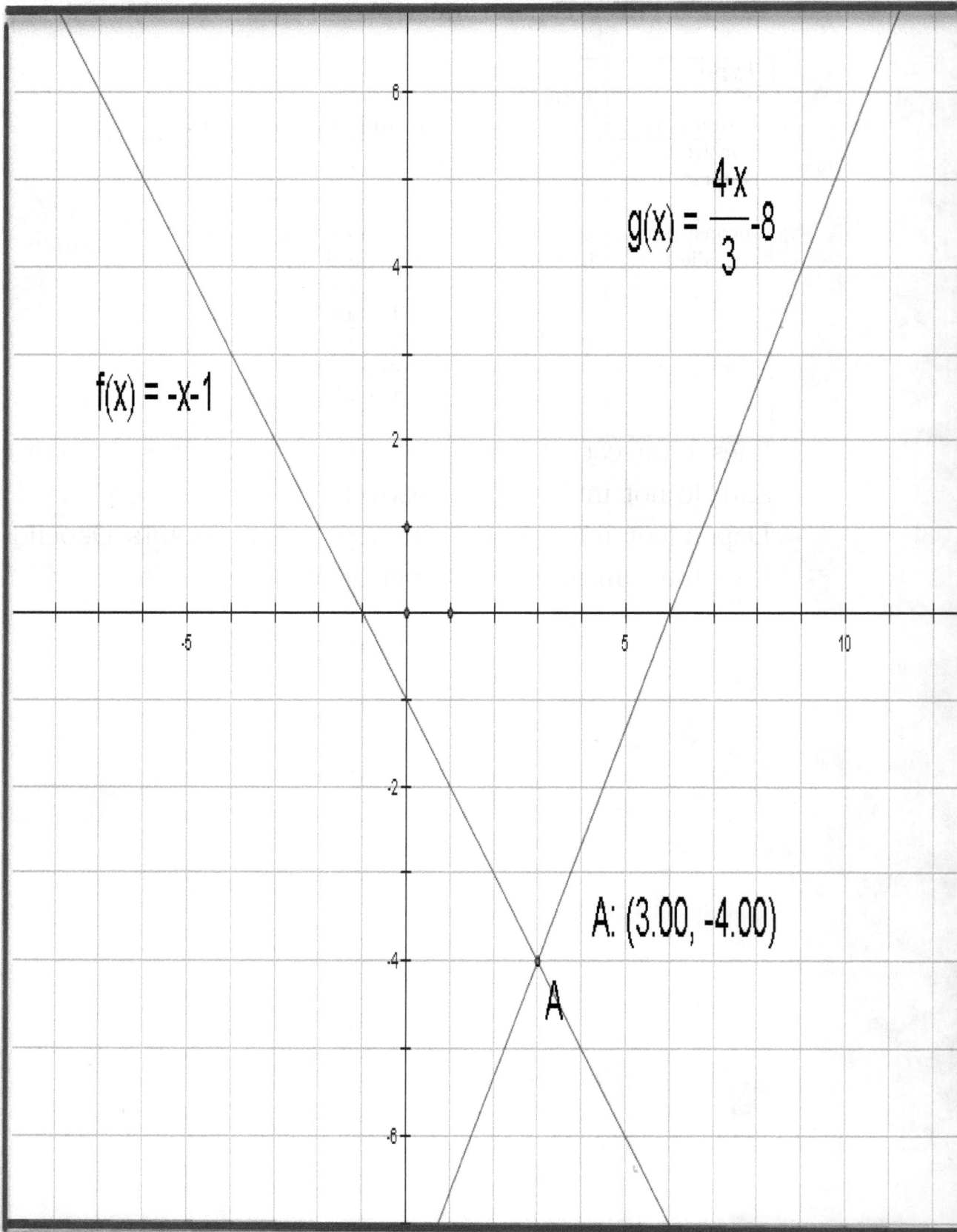

$$g(x) = \frac{4 \cdot x}{3} - 8$$

$$f(x) = -x - 1$$

A: (3.00, -4.00)

A

Figure 12.4: Intersecting lines in a linear system

$$f(X) = \frac{-3 \cdot X}{-2} + \frac{6}{-2}$$

$$g(X) = \frac{-6 \cdot X}{-4} + \frac{18}{-4}$$

parallel lines

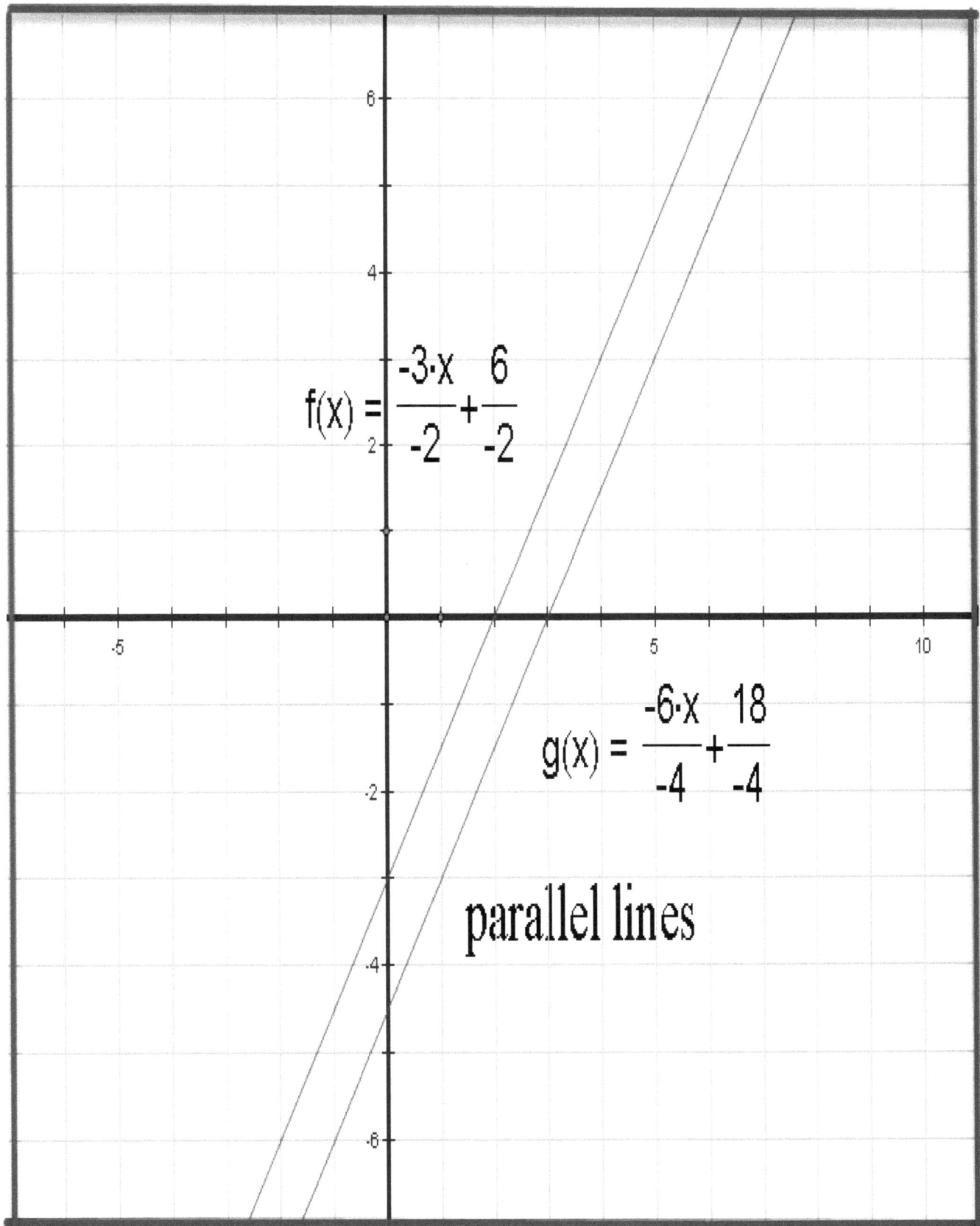

Figure 12.5: Parallel lines in a linear system

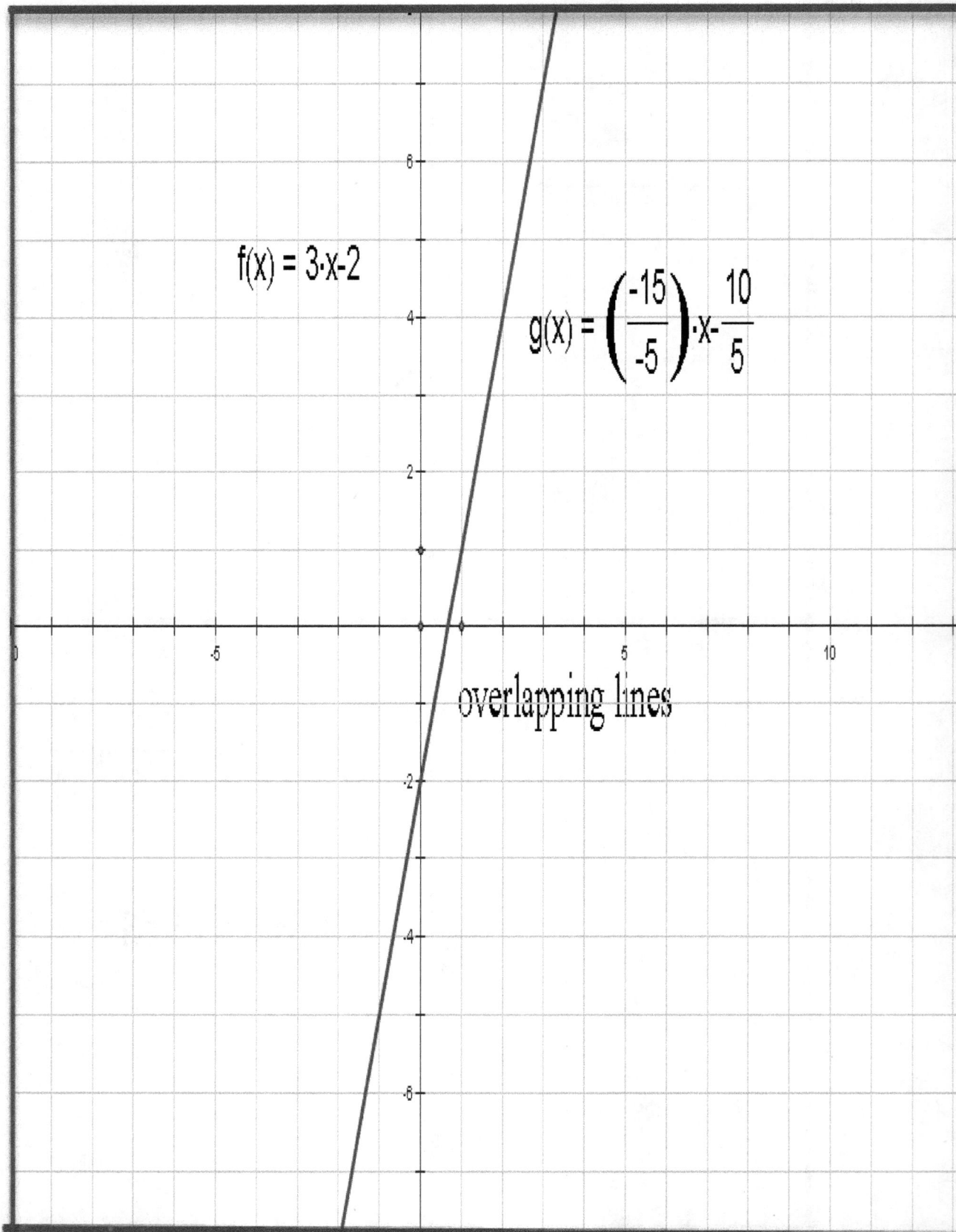

$$f(x) = 3 \cdot x - 2$$

$$g(x) = \left(\frac{-15}{-5}\right) \cdot x - \frac{10}{5}$$

overlapping lines

Figure 12.6: Overlapping lines in a linear system

Three-types-Linear-Solutions Definition: A system is when 2 or more equation
are defined together,
A graphic example solution would be
when the two equations intersect. **Consistant System**

$$y = -x - 1$$

$$4x - 3y = 24$$

Rearrange and solve for y

$$y = -x - 1$$

$$-3y = -4x + 24$$

Multiply the first equation by 3, then add the two equations.

$$3y = -3x - 3$$

$$-3y = -4x + 24$$

$$0 = -7x + 21$$

$$7x = 21$$

$$x = 3$$

$$y = -x - 1$$

$$y = -(3) - 1$$

$$y = -4$$

$$\text{Intersecting Point} = \{3, -4\}$$

Inonsistant System

$$3x - 2y = 6$$

$$6x - 4y = 18$$

$$3x - 2y = 6 \text{ Multiply by } -2$$

$$-6x + 4y = -12$$

$$6x - 4y = 18$$

$$0x - 0y = 6$$

$$0 = 6 \quad \text{False}$$

$$m_1 = m_2 : \text{The slopes are equal}$$

and the lines are parallel

The two lines do not intersect

like railroad tracks. (Hopefully)

Dependent System

$$y = 3x - 2$$

$$15x - 5y = 10$$

$$-3x + y = -2 \quad \text{Multiply by 5}$$

$$-15x + 5y = -10$$

$$15x - 5y = 10$$

$$0 = 0 \quad \text{True}$$

Chapter 13

Solving 3 vars. x 3 unkn.

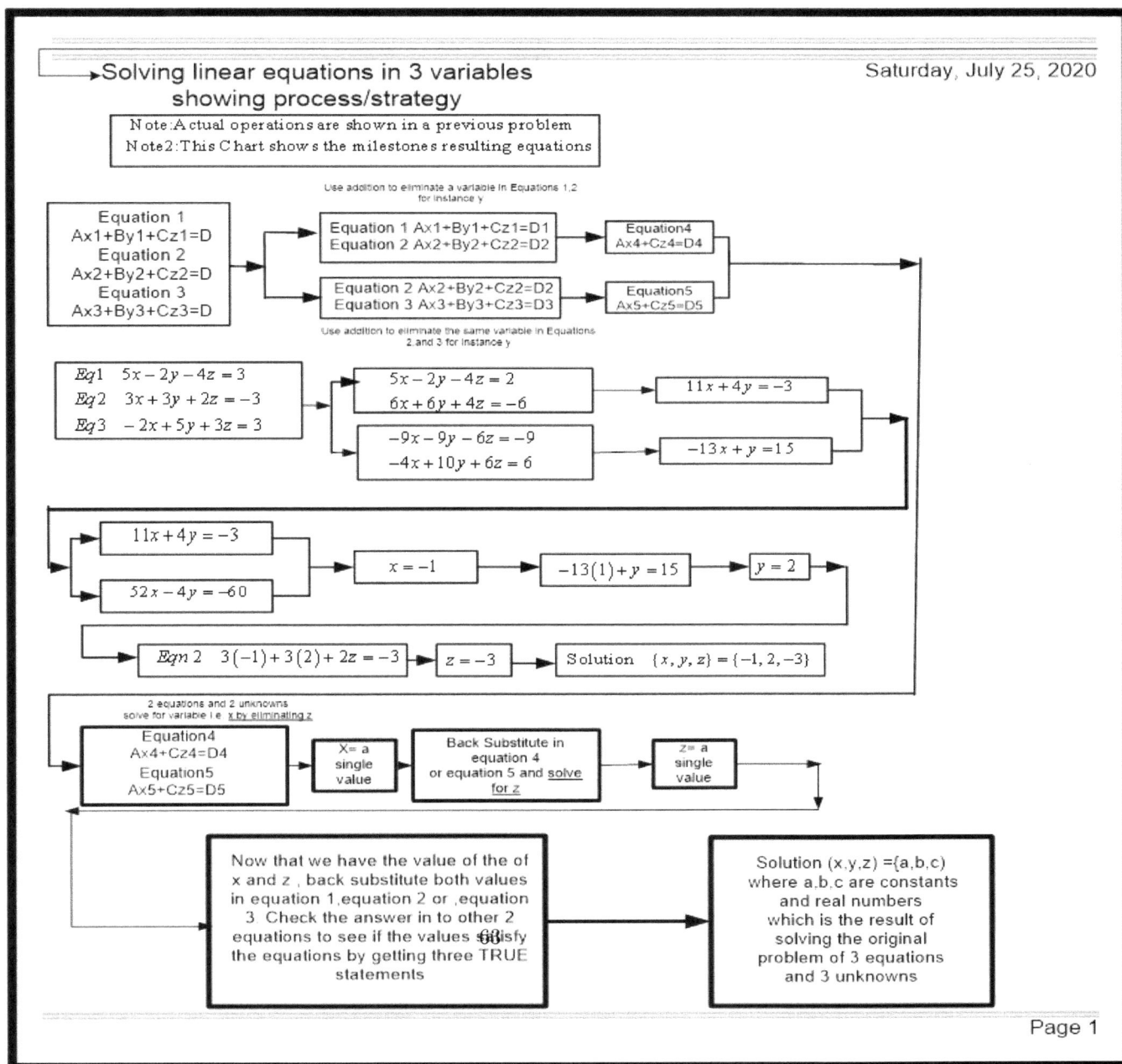

Note: Actual operations are shown in a previous problem
Note2: This Chart shows the milestones resulting equations

Use addition to eliminate a variable in Equations 1,2 for instance y

Equation 1
$Ax1+By1+Cz1=D$
Equation 2
$Ax2+By2+Cz2=D$
Equation 3
$Ax3+By3+Cz3=D$

Equation 1 $Ax1+By1+Cz1=D1$
Equation 2 $Ax2+By2+Cz2=D2$

Equation4
$Ax4+Cz4=D4$

Equation 2 $Ax2+By2+Cz2=D2$
Equation 3 $Ax3+By3+Cz3=D3$

Equation5
$Ax5+Cz5=D5$

Use addition to eliminate the same variable in Equations 2 and 3 for instance y

$Eq1 \quad 5x - 2y - 4z = 3$
$Eq2 \quad 3x + 3y + 2z = -3$
$Eq3 \quad -2x + 5y + 3z = 3$

$5x - 2y - 4z = 2$
$6x + 6y + 4z = -6$

$11x + 4y = -3$

$-9x - 9y - 6z = -9$
$-4x + 10y + 6z = 6$

$-13x + y = 15$

$11x + 4y = -3$
$52x - 4y = -60$

$x = -1$

$-13(1) + y = 15$

$y = 2$

$Eqn 2 \quad 3(-1) + 3(2) + 2z = -3$

$z = -3$

Solution $\{x, y, z\} = \{-1, 2, -3\}$

2 equations and 2 unknowns solve for variable i.e. x by eliminating z

Equation4
$Ax4+Cz4=D4$
Equation5
$Ax5+Cz5=D5$

$X = a$ single value

Back Substitute in equation 4
or equation 5 and solve for z

$z = a$ single value

Now that we have the value of the of x and z , back substitute both values in equation 1, equation 2 or , equation 3. Check the answer in to other 2 equations to see if the values satisfy the equations by getting three TRUE statements

Solution $(x,y,z) = \{a,b,c\}$ where a.b.c are constants and real numbers which is the result of solving the original problem of 3 equations and 3 unknowns

13.1   3 equations and 3 unknowns

3 equations and 3 unknowns, elimination by addition,with the result [x,y,z]=[-1,2,-3]

13.2   General Format for 3 equations and 3 variables

General Format

$$Ax_1 + By_1 + Cz_1 = D_1$$
$$Ax_2 + By_2 + Cz_2 = D_2$$
$$Ax_3 + By_3 + Cz_3 = D_3$$

13.3   System to be solved 3 variables 3 equations

13.3.1   Reduce the system to 2 equations and 2 variables.

Method using simultaneous equations,

13.3.2   Eliminate variable z

More detail

Equation 1: $5x - 2y - 4z = 3$

Equation 2: $3x + 3y + 2z = -3$

Equation 3: $-2x + 5y + 3z = 3$

Reduce the system to 2 equations and 2 variables.

Equation 1: $5x - 2y - 4z = 3$ No change

Equation 2: $3x + 3y + 2z = -3$ Multiply by 2

Equation 1: $5x - 2y - 4z = 3$

Equation 2: $6x + 6y + 4z = -6$

13.4 Add the resulting equations to eliminate z,

Add the resulting equations to eliminate z,

and add two remaining equations to produce "Equation 4".

Equation 4: $11x + 4y = -3$

13.5   Reduce the system to 2 equations and 2 variables

13.5.1   Eliminate variable z in a second set of equations.

Reduce the system to 2 equations and 2 variables

This time use Equation 2 and Equation 3.

Follow the same process as above. We eliminate the same variable
  z .

Choose Equation 2 and Equation 3

Equation 2: $3x + 3y + 2z = -3$ Multiply by -3

Equation 3: $-2x + 5y + 3z = 3$ Multiply by 2

Multiply thru to get new equivalent equations

Equation 2b: $-9x + -9y - 6z = 9$

Equation 3b: $-4x + 10y + 6z = 6$

Add Equation 2b and 3b to produce Equation 5

Equation 5: $-13x + y = 15$

13.6   Add equations 4 and equation 5 to eliminate y, and solve for x.

Add equations 4 and equation 5 to eliminate y

Equation 4: $11x + 4y = -3$ No change

Equation 5: $-13x + y = 15$ Multiply by -4

Equation 4b: $11x + 4y = -3$

Equation 5b: $52x - 4y = -60$

13.7   Result of adding equation 4b and equation 5b

63x=-63

x=-1

13.8   Back substitute value of x in equation 5

Equation 5: $-13x + y = 15$

Equation 5: $-13(-1) + y = 15$

Equation 5: $13 + y = 15$

$y = 2$

13.9   Back substitute the values of x and y in equation 2 to solve for z

Equation 2: $3x + 3y + 2z = -3$

Equation 2: $3(-1) + 3(2) + 2z = -3$

Equation 2: $3 + 2z = -3$

2z=-6

z=-2

13.9.1   Back substitute x,y and z in equation 2, check the solution

13.9.2   Equation 1 or 2 or 3 can be used to verify the solution

The solution is (x, y, z)=(-1, 2, -3)

Test answer using equation 1

Equation 1: $5x - 2y - 4z = 3$

5(-1)-2(2)-3(-3)=3

-9+12=3

3=3 It checks out.

Chapter 14

Solving a 2x2 linear system Using Cramers rule

14.1 Explanation and setup of Cramers Rule

Cramer's rule is a method to solve a linear equations system by using the coefficients of the x, y, c in a two-equation system and x, y, z, c in a three-equation system

Definition-1: An augmented matrix is a matrix that has the x, y, and constant in an ordered matrix.

Definition-2: A determinate D uses the x and y coefficients

Definition-3: A determinate Dx used the y and constant values

Definition-4: A determinate Dy used the x and constant values

Where the constant value is on the right side of the equals sign

Note-1: The D uses the values on the left side of the equation.

Note-2: The Dx uses the remainders of the columns not x.

Note-3: The Dy uses the remainders of the columns not y.

$$D = \begin{pmatrix} x_1 & y_1 \\ x_2 & y_2 \end{pmatrix}$$

$$D_x = \begin{pmatrix} c_1 & y_1 \\ c_2 & y_2 \end{pmatrix}$$

$$D_y = \begin{pmatrix} x_1 & c_1 \\ x_2 & c_2 \end{pmatrix}$$

The determinate of a 2x2 matrix is

$$\begin{bmatrix} a_1 & b_1 \\ a_2 & b_2 \end{bmatrix}$$

And is defined as

$$D = \begin{bmatrix} a_1 & b_1 \\ a_2 & b_2 \end{bmatrix}$$

Its value is the difference cross product of its values starting with

$$a_1 b_2 - a_2 b_1$$

Therefore:

$$D = \begin{bmatrix} a_1 & b_1 \\ a_2 & b_2 \end{bmatrix} = (a_1 b_2 - a_2 b_1)$$

The augmented matrix is used to solve two linear equations in a system of two equations.

$$x = \frac{D_x}{D} \text{ and } y = \frac{D_y}{D}$$

The second solution involves using Cramer's rule. The determinate of a 2 x 2 is:

$$\begin{bmatrix} a_1 & b_1 \\ a_2 & b_2 \end{bmatrix} \text{ and is defined by } \quad D = \begin{bmatrix} a_1 & b_1 \\ a_2 & b_2 \end{bmatrix}$$

and the values needed are:

$$D = \begin{bmatrix} a_1 & b_1 \\ a_2 & b_2 \end{bmatrix}$$

$$D_x = \begin{bmatrix} c_1 & y_1 \\ c_2 & y_2 \end{bmatrix}$$

$$D_y = \begin{bmatrix} x_1 & c_1 \\ x_2 & c_2 \end{bmatrix}$$

14.2   Cramer's solution

14.2.1   Two equations system, with the result 8,-3

Solve the system of two equations by Cramer's Rule

Solve the system of two equations by Cramer's Rule

$$3x + 11y = -9$$
$$7x + 6y = 38$$

$$\begin{vmatrix} x_1 + y_1 = c_1 \\ x_2 + y_2 = c_2 \end{vmatrix}$$

Use the above format

$$\begin{vmatrix} 3x + 11y = -9 \\ 7x + 6y = 38 \end{vmatrix}$$

$$x = \frac{D_x}{D} = \frac{\begin{bmatrix} c_1 & b_1 \\ c_2 & b_2 \end{bmatrix}}{\begin{bmatrix} a_1 & b_1 \\ a_2 & b_2 \end{bmatrix}} = \frac{(c_1 b_2 - c_2 b_1)}{(a_1 b_2 - a_2 b_1)}$$

$$x = \frac{(c_1 b_2 - c_2 b_1)}{(a_1 b_2 - a_2 b_1)} = \frac{((-9*6)-(38*11))}{((3*6)-(7*11))} = 8$$

$$x = = \frac{(-472)}{(-59)} = 8$$

$$y = \frac{D_y}{D} = \frac{\begin{bmatrix} a_1 & c_1 \\ a_2 & c_2 \end{bmatrix}}{\begin{bmatrix} a_1 & b_1 \\ a_2 & b_2 \end{bmatrix}} = \frac{(a_1 c_2 - a_2 c_1)}{(a_1 b_2 - a_2 b_1)}$$

$$y = \frac{(a_1 c_2 - a_2 c_1)}{(a_1 b_2 - a_2 b_1)} = \frac{((3*38)-(7*9))}{((3*6)-(7*11))} = -3$$

$$y = = \frac{(177)}{(-59)} = -3$$

Chapter 15

3 equations and 3 unknowns solved by Cramer's rule

15.0.1 3x3 General Formula

General Format

*Equation*1	$Ax_1 + By_1 + Cz_1 = D_1$
*Equation*2	$Ax_2 + By_2 + Cz_2 = D_2$
*Equation*3	$Ax_3 + By_3 + Cz_3 = D_3$

System to be solved

*Equation*1	$5x - 2y - 4z = 3$
*Equation*2	$3x + 3y + 2z = -3$
*Equation*3	$-2x + 3y + 2z = -3$

15.1 3x3 system solution using determinates

D is the determinate for the coefficients on the left side of the equation set.

D_x Replace the x column with the right side of the equation system.

D_y Replace the y column with the right side of the equation system.

D_z Replace the z column with the right side of the equation system.

15.2   solve for x, y, z, using determinates

$$x = \frac{D_x}{D}$$

$$y = \frac{D_y}{D}$$

$$z = \frac{D_z}{D}$$

15.3   In general, a 2 by 2 matrix is solved using this formula

In general, a 2 by 2 matrix is solved using this formula

$$D_{2x2} = \begin{bmatrix} a_1 & b_1 \\ a_2 & b_2 \end{bmatrix} = (a_1 b_2 - a_2 b_1)$$

15.4   In general, a 3 by 3 matrix is solved using this formula

In general, a 3 by 3 matrix is solved using this formula.
Notice that the operators independent
of the values used are $+, -, +$
In other words, the second operator is a minus [-] independent
of the of the values used in a particular problem.

$$D_{3x3} = \begin{vmatrix} a_1 & b_1 & c_1 \\ a_2 & b_2 & c_2 \\ a_3 & b_3 & c_3 \end{vmatrix} = a_1 \begin{bmatrix} b_2 & c_2 \\ b_3 & c_3 \end{bmatrix} - a_2 \begin{bmatrix} b_1 & c_1 \\ b_3 & c_3 \end{bmatrix} + a_3 \begin{bmatrix} b_1 & c_1 \\ b_2 & c_2 \end{bmatrix}$$

15.5 Specific 3 x 3 determinate for Cramer's solution.

Specific 3 x 3 determinate for Cramer's solution.

$$D = \begin{vmatrix} a_1 & b_1 & c_1 \\ a_2 & b_2 & c_2 \\ a_3 & b_3 & c_3 \end{vmatrix} = a_1 \begin{bmatrix} b_2 & c_2 \\ b_3 & c_3 \end{bmatrix} - a_2 \begin{bmatrix} b_1 & c_1 \\ b_3 & c_3 \end{bmatrix} + a_3 \begin{bmatrix} b_1 & c_1 \\ b_2 & c_2 \end{bmatrix}$$

Equation 1 $5x - 2y - 4z = 3$
Equation 2 $3x + 3y + 2z = -3$
Equation 3 $3x + 3y + 2z = -3$

Column $\vert$ x y $z\vert$ c $\vert$

$$D = \begin{vmatrix} 5 & -2 & -4\vert & 3 \vert \\ 3 & 3 & 2\vert & -3\vert \\ -2 & 5 & 3\vert & 3\vert \end{vmatrix}$$

15.6 Resolved matrix using the coefficients

Resolved matrix using the coefficients
of the constant in the x, y, z,
columns for D, D_x, D_y, D_z

Calculate D

$$D = \begin{vmatrix} 5 & -2 & -4\vert \\ 3 & 3 & 2 \\ -2 & 5 & 3 \end{vmatrix} =$$

$$D = 5 \begin{bmatrix} 3 & 2 \\ 5 & 3 \end{bmatrix} - \begin{bmatrix} -2 & -4 \\ 5 & 3 \end{bmatrix} + (-2) \begin{bmatrix} -2 & -4 \\ 3 & 2 \end{bmatrix} =$$

$3[(3 \bullet 3) - (5 \bullet 2)] \ + 3[(-2 \bullet 3) -$
$(5 \bullet (-4))] \ + (-2) [((-2) \bullet 2) - (3 \bullet 4)]$
$5[(9 - 10] \quad - 3[-6 + 20] \quad + (-2) [(-4) - (-12)]$

$$(-5) \quad + (-42) \quad + (-16) = -63 = D$$

$$D = -63$$

Calculate D_x

$$D_x = \begin{vmatrix} 3 & -2 & -4 \\ -3 & 3 & 2 \\ 3 & 5 & 3 \end{vmatrix}$$

$$D_x = 3 \begin{bmatrix} 3 & 2 \\ 5 & 3 \end{bmatrix} - (-3) \begin{bmatrix} -2 & -4 \\ 5 & 3 \end{bmatrix} + 3 \begin{bmatrix} -2 & -4 \\ 3 & 2 \end{bmatrix}$$

$$3[(3 \bullet 3) - (5 \bullet 2)] \ + 3[(-2 \bullet 3) -$$
$$(5 \bullet (-4))] \ + (3) [((-2) \bullet 2) - (3 \bullet 4)]$$
$$3(-1) + 3(14) + 3(8) = -3 + 42 + 24 = 63 = D_x$$

$$D_x = 63$$

Calculate D_y

$$D_y = \begin{vmatrix} 5 & 3 & -4 \\ 3 & -3 & 2 \\ -2 & 3 & 3 \end{vmatrix}$$

$$D_y = 5 \begin{bmatrix} -3 & 2 \\ 3 & 3 \end{bmatrix} - 3 \begin{bmatrix} 3 & -4 \\ 3 & 3 \end{bmatrix} + (-2) \begin{bmatrix} 3 & -4 \\ -3 & 2 \end{bmatrix}$$

$$5[(-3 \bullet 3) - (3 \bullet 2)] \ + 3[(3 \bullet 3) - (3 \bullet (-4))] \ + (-2) [(3 \bullet 2) - (-3 \bullet -4)]$$

$$5(-9 - 6) \quad - 3(9 - (-12)) \quad + (-2)(6 - 12)$$

$$-75 - 63 + 12 = -126 = D_y$$

$$D_y = -126$$

Calculate D_z

$$D_z = \begin{vmatrix} 5 & -2 & 3 \\ 3 & 3 & -3 \\ -2 & 5 & 3 \end{vmatrix}$$

$5[(3 \bullet 3) - (5(-3))] - 3[(-2 \bullet 3) - (5 \bullet 3)] + (-2)[(-2 \bullet -3) - (3 \bullet 3)]$

$5[24] - 3[-21] + (-2)[-3]$

$120 + 63 + 6 = 189 = D_z$

$D_z = 189$

15.7 Solution to determinate problem

$$x = \frac{D_x}{D}$$

$$y = \frac{D_y}{D}$$

$$z = \frac{D_z}{D}$$

$$x = \frac{D_x}{D} = \frac{63}{-63} = -1$$

$$y = \frac{D_y}{D} = \frac{-126}{-63} = 2$$

$$z = \frac{D_z}{D} = \frac{189}{-63} = -3$$

The solution is (x, y, z,)=(-1,2,-3)

Test answer using equation 1

Equation 1:

$$5x - 2y - 4z = 3$$

5(-1)-2(2)-3(-3)=3

-9+12=3

3=3 It checks out.

Chapter 16

Inequalities and Absolute value

$|x| = c$ Absolute value Equals c

$|x| \geq c$ Absolute value Greater than or equal to c

$|x| > c$ Absolute value Greater than c

$|x| \leq c$ Absolute value Less than or equal to c

$|x| < c$ Absolute value Less than c

The arrow points to the smaller value.

16.1 Absolute Values

16.2 Inequalities and Equals on Real Number Line.

- Inequalities are Ranges whereas Equals are Points on a real number line

- Define x, absolute x and the resultant c and graph on a real number line.

- Absolute values are shown bounded by two vertical lines $|x|$ with the result in this general case represented by c

- Since x is within the absolute value brackets $|x|$, x can be either positive or negative, and c is always positive. C represents any real number.

- The result of an absolute value is always positive therefore: $|-x| = c$ and $|x| = c$

Subject: Absolute Values shown as Inequalities and Equalities on Real Number Line..

Define x, absolute x and the resultant c and graph on a real number line.

Note: Inequalities are Ranges whereas Equalities are Points on a real number line

$|x| = c$ Absolute values are shown bounded by two vertical lines | x | with the result in this general case represented by c

Since x is within the absolute value brackets $|x|$, x can be either positive or negative, and c is always positive. C represents any real number.

The result of an absolute value is always positive therefore: |-x|=c and |x|=c

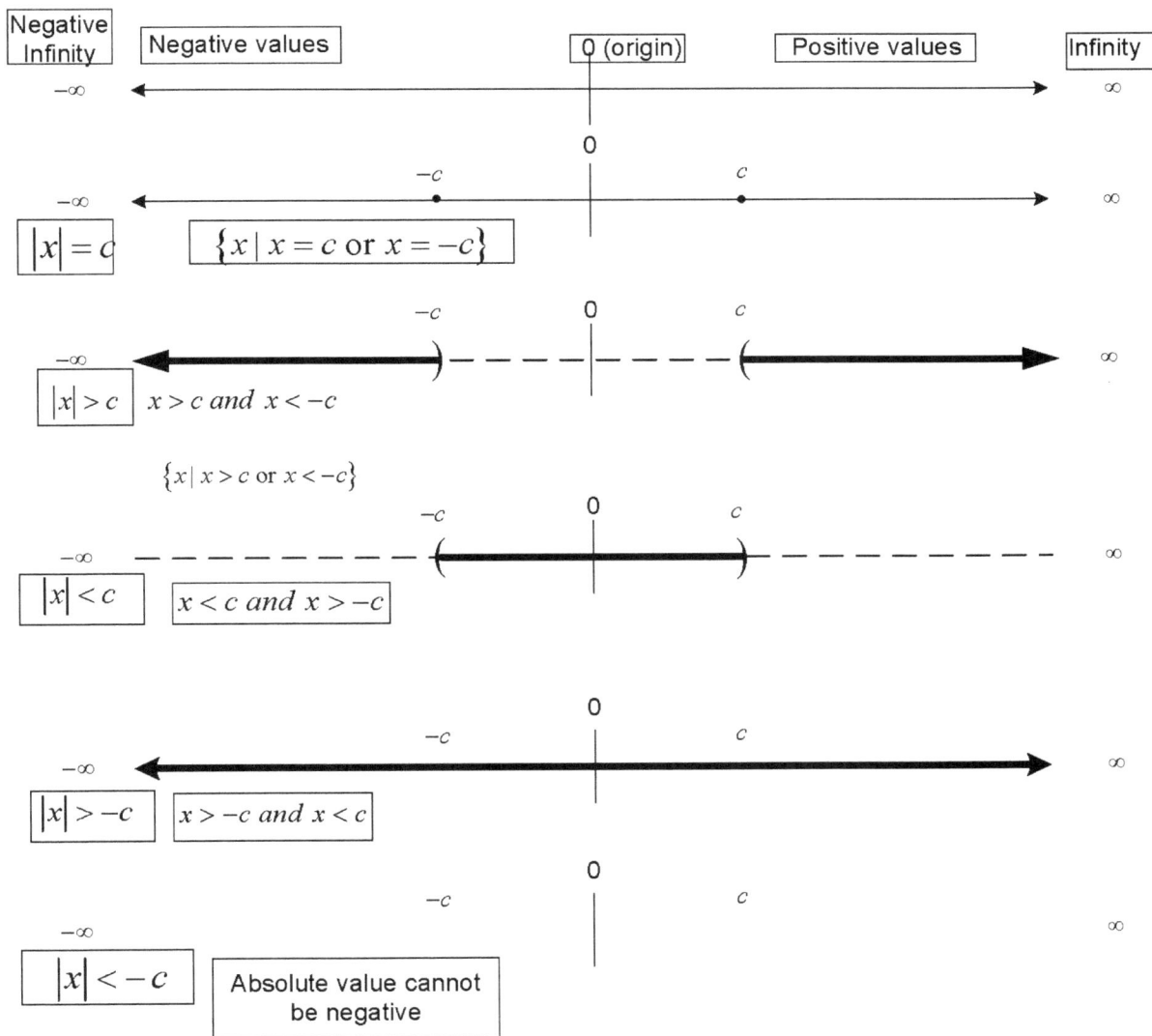

| Negative Infinity | Negative values | | 0 (origin) | | Positive values | Infinity |

$-\infty$ ⟵————————————————————————⟶ ∞

0

$-c$ c

$-\infty$ ⟵————•————————•————————⟶ ∞

$|x| = c$ $\{x \mid x = c \text{ or } x = -c\}$

$-c$ 0 c

$-\infty$ ⟵━━━━━) – – – | – – – (━━━━━⟶ ∞

$|x| > c$ $x > c$ and $x < -c$

$\{x \mid x > c \text{ or } x < -c\}$

$-c$ 0 c

$-\infty$ – – – (━━━━━|━━━━━) – – – ∞

$|x| < c$ $x < c$ and $x > -c$

0

$-c$ c

$-\infty$ ⟵━━━━━━━━━━━━|━━━━━━━━━━━━⟶ ∞

$|x| > -c$ $x > -c$ and $x < c$

0

$-c$ c

$-\infty$ | ∞

$|x| < -c$ Absolute value cannot be negative

Figure 16.1: Absolute-Value-inequality-chart

Chapter 17

Absolute Value and Inequality calculations

- Absolute-Value-inequality calculations for $|3 - 2x| > 7$
- Remove the absolute value bars $|(\text{function})|$
- $|3 - 2x| > 7$ Means $3 - 2x > 7$ and $3 - 2x < -7$
- There are two solutions to an absolute value expression
- Solution 1 : $3 - 2x > 7$ No Change
- Solution 2 : $3 - 2x > 7$ Multiply by -1 to get absolute value
- $(-1)(3 - 2x > 7) = \ -(3 - 2x) < (-7) =$
- Solution 2: $-3 + 2x < -7$
- Solution 1: $3 - 2x = 7; x = -2$; Left intersecting point
- Solution 2: $-(3 - 2x) = 7; x = 5$; Right intersecting point
- Note the change of direction of the in-equals sign
- Multiplying (or dividing) by -1 (minus 1) changes the direction if the inequality sense.
- In this case multiplying by -1 changed the sign from "greater than" " $>$ " to "less than" $<$ ".
- Given two solutions both solutions are now resolved.

17.1   The solutions equal the projection on the real number line.

- The solutions represent the projection on the real number line, of intersection of the absolute value equation , and the "y" value it tested against.

- Solve for the projection of the intersection of y=7 and

$$|3 - 2x|$$

  by setting each side of the equation to 7 and solve for x.

- Solution 1: $3 - 2x = 7; x = -2;$ Left intersecting point

- Solution 2: $-(3 - 2x) = 7; x = 5;$ Right intersecting point

17.2   absolute value equations

Subject: Absolute Values shown as Inequalities and Equalities on Real Number Line..

Define x, absolute x and the resultant c and graph on a real number line.

Note: Inequalities are Ranges whereas Equalities are Points on a real number line

$|x| = c$ Absolute values are shown bounded by two vertical lines | x | with the result in this general case represented by c

Since x is within the absolute value brackets $|x|$, x can be either positive or negative, and c is always positive. C represents any real number.

The result of an absolute value is always positive therefore: |-x|=c and |x|=c

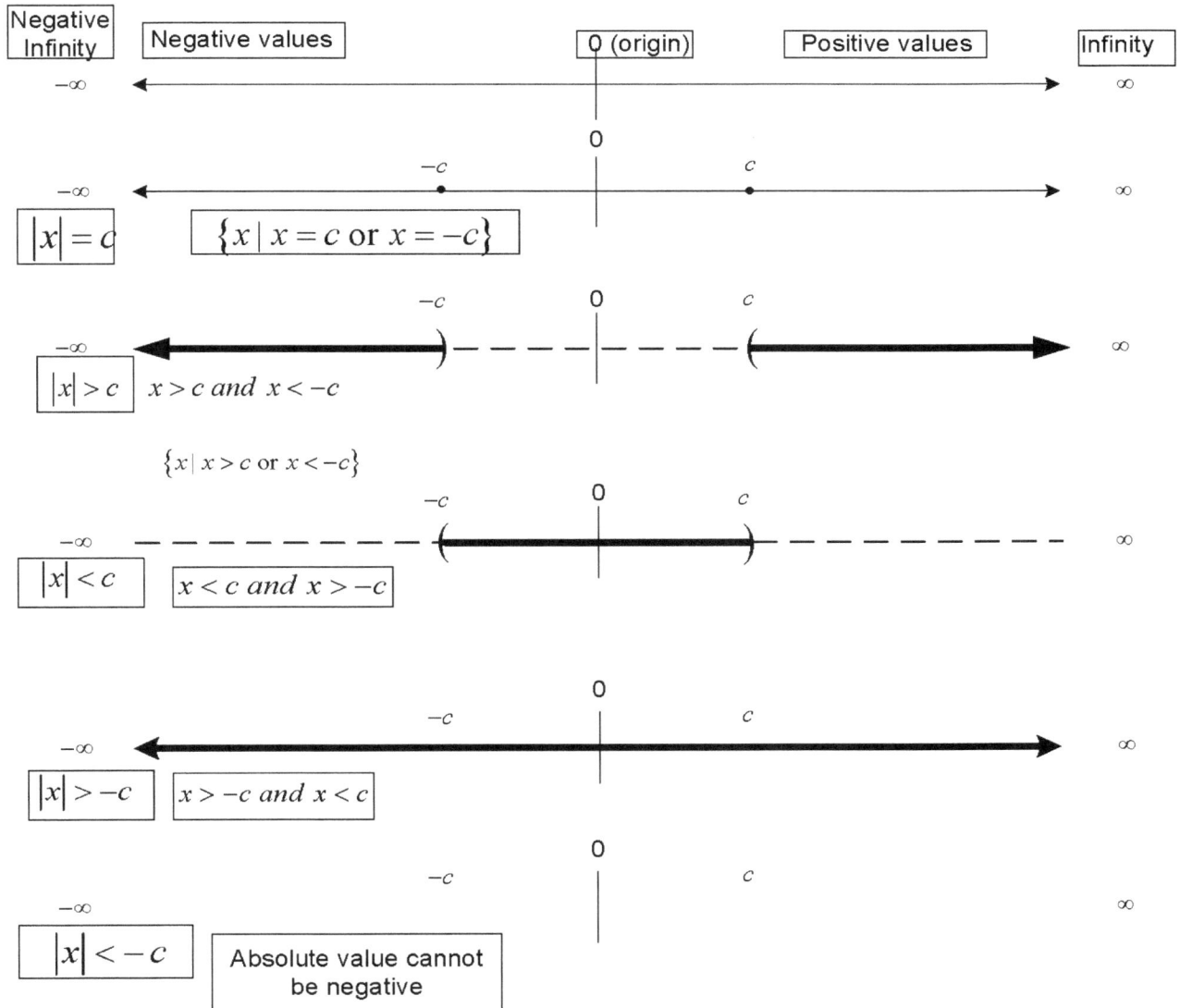

Negative Infinity   Negative values   0 (origin)   Positive values   Infinity

$-\infty$ $\longleftrightarrow$ ∞

0

$-c$ c

$-\infty$ $\longleftrightarrow$ ∞

$|x| = c$ $\{x \mid x = c \text{ or } x = -c\}$

$-c$ 0 c

$-\infty$ $\longleftrightarrow$ ∞

$|x| > c$ $x > c \text{ and } x < -c$

$\{x \mid x > c \text{ or } x < -c\}$

$-c$ 0 c

$-\infty$ ∞

$|x| < c$ $x < c \text{ and } x > -c$

0

$-c$ c

$-\infty$ $\longleftrightarrow$ ∞

$|x| > -c$ $x > -c \text{ and } x < c$

0

$-c$ c

$-\infty$ ∞

$|x| < -c$ Absolute value cannot be negative

Negative Infinity　　Negative values　　　　　　　　　　0 (origin)　　　　　　　　Infinity

$-\infty$　　　　　　　　　　　　　　　　　　　　　　　　　∞

$-c$　　　　　0　　　　c

$-\infty$　　　　　　　　　　　　　　　　　　　　　∞

$|x| > c$　　$\{\,x \mid x < -c \quad \text{or} \quad x > c\,\}$

Negative values　　　　　　　　0 (origin) →　　　Positive values

-11 -10 -9 -8 -7 -6 -5 -4 -3 -2 -1 0 1 2 3 4 5 6 7 8 9 10 11

$-\infty$　　　　　　　　　　　　　　　　　　　　　　　　　　∞

$f(x) = |3 - 2x| > 7$

$f(x) = |3 - 2 \cdot x|$　　　　　　　　　$r(x) = |3 - 2 \cdot x|$

$h(x) = 7$

$g(y) = -2$　　　　　　　　　$q(y) = 5$

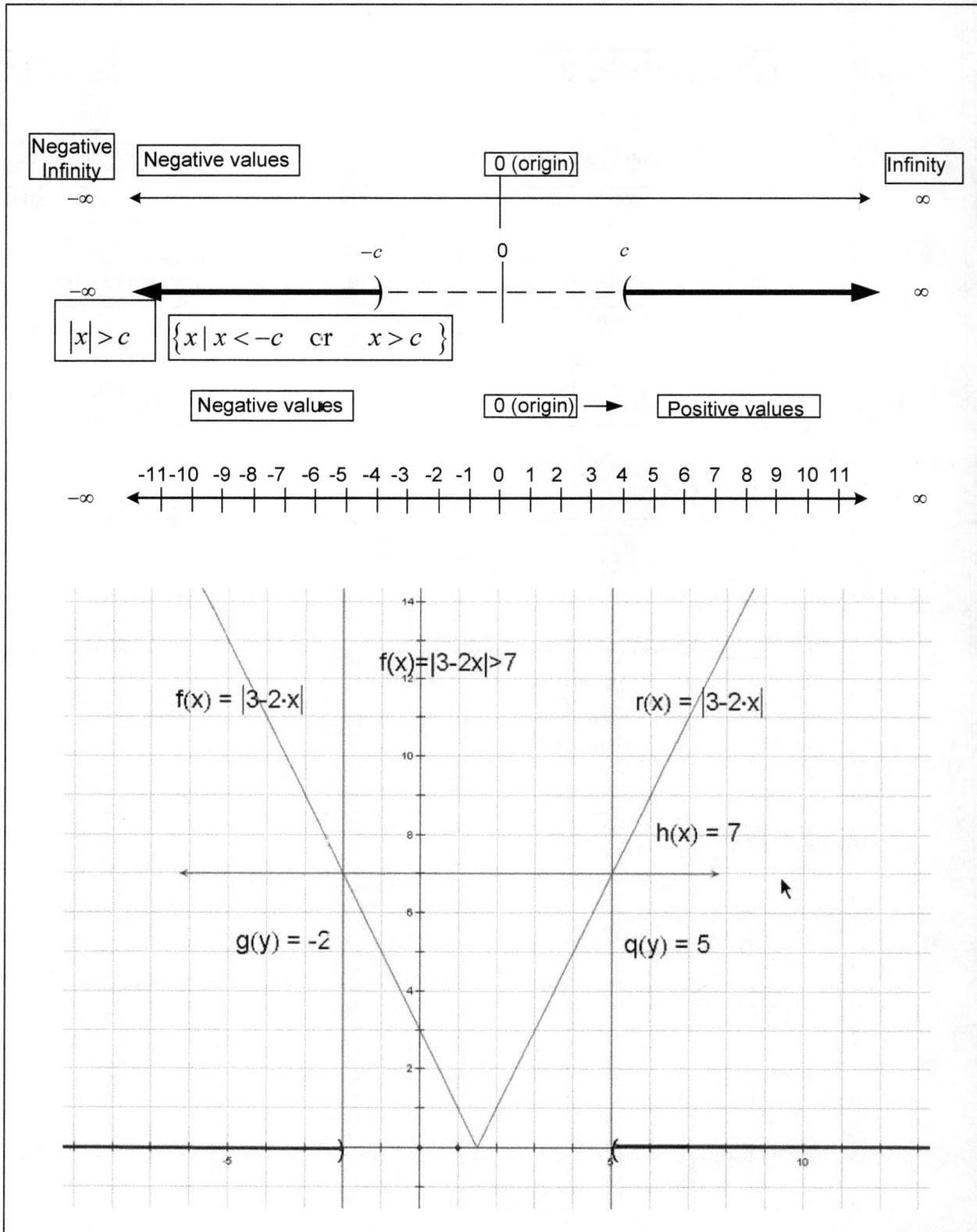

Figure 17.2: Absolute-Value-inequality

Chapter 18

Applied Algebra

1. Step 1 **Read the problem carefully**
   until you understand what is given and what is to be found.
2. Step 2 **Assign a variable** to represent the unknown value.
   (a) Use diagrams or tables as needed.
   (b) Write down what the variable represents.
   (c) Then, express any other unknown values in terms of the first variable.
3. Step 3 **Write an equation** using the variable expression(s).
4. Step 4 **Solve the equation.**
5. Step 5 **State the answer.**
6. Step 6 **Check the answer** in the words of the original problem.
   (a) Does the solution seem reasonable?
7. Step 6 **Check the answer** in the words of the original problem.

18.1   Garden Problem Dimensions

Garden Problem Dimensions

$\leftarrow Width = (x-4) \longrightarrow$

$Length = x$

Garden

$Area = 96\,feet^2$

$Area = Length * Width$

$Length = x$

$Width = (x-4)$

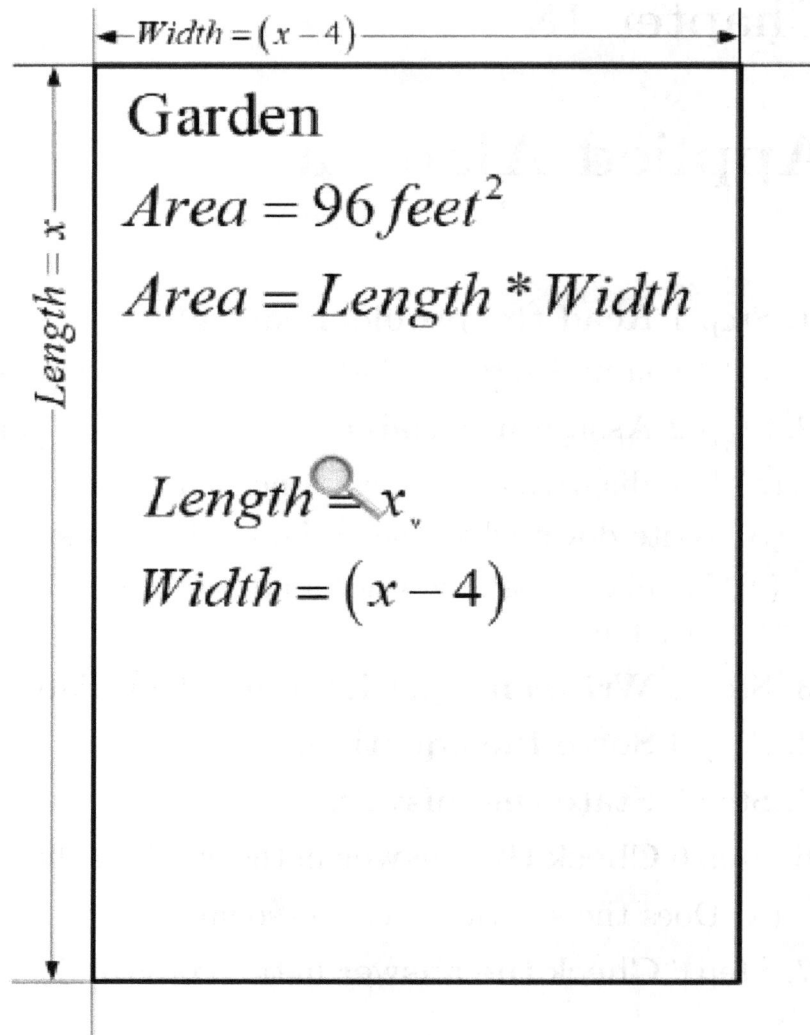

Solution: See Solution Sheet calculations.

$Length = x = 12\,ft$

$Width = (x-4)\,ft = (12-4)\,ft = 8\,ft$

18.2 Applied Algebra

1. Step 1 **Read the problem carefully** until you understand what is given and what is to be found.
   The width of the garden will be 4 feet less than the garden's length.
   We need to find the dimensions of a garden with area 96 ft^2.

2. Step 2 **Assign a variable** to represent
   the unknown value, using diagrams or
   tables as needed. Write down what the variable represents.
   Then, express any other unknown values in terms of the first variable.
   Let x = the length of the garden. Then x - 4 = the width.
   (The width is 4 ft less than the length.)

3. Step 3 **Write an equation** using the variable expression(s).
   Area=Length times Width $A = LW$ therefore $96 = x(x - 4)$

4. Step 4 **Solve the equation.** Formula $96 = x(x - 4)$
   Multiply and solve for Zero $0 = x^2 - 4x - 96$
   Factor $(x - 12)(x + 8) = 0$
   Set each factor equal to zero and solve $(x - 12) = 0$ Then $x = 12$
   Set remaining factor equal to zero and solve $(x + 8) = 0$
   Then $x = -8$

5. Step 5 **State the answer.**
   Does it seem reasonable?
   Given that the answers are -8 and 12 we discard the -8 since dimensions are positive.
   We accept the 12 answer since it is positive.

6. Step 6
   Check the answer in the words of the original problem.
   If $x = 12$ then $(x - 4) = (12 - 4) = 8$

Then the dimensions are

Width=8 and Length =12.

If A=lw then 96=(8)(12)=96 so the answer checks out.

Steps $1 - 6$

Math Solution in an Algebra formulas progression.

Solution

$$A = LW$$

$$96 = x(x - 4)$$
$$0 = x^2 - 4x - 96$$
$$(x - 12)(x + 8) = 0$$
$$(x - 12) = 0$$
$$x = 12$$
$$(x + 8) = 0$$

Not used x=-8

$$x = -8$$

Accepted x=12

$$x = 12$$

Check:

$$(x - 4) = (12 - 4) = 8$$
$$A - LW$$
$$96 = (8)(12) = 96$$

18.3 More Problem Solving in Algebra-1

18.3.1 Calculate two-person time to complete 1 spraying job.

Sam and Chester can complete a spraying job together but at different rates

1. Sam takes 8 hours to do the work alone

2. Sam's Rate per hour is 1/8 of the job in one hour

3. Chester takes 14 hours to do the work alone

4. Chester's Rate per hour is 1/14 of the job in one hour

5. The time working together is x the same for both Sam and Chester

Worker	Rate	Working Together Time	Factional Part of Work Done
Sam	1/8	x	1/8 x
Chester	1/14	x	1/14 x

1. Rate per hour equals work done in one hour.

2. Time equals x

3. Work equals rate * time

4. Work in entire job equals 1

5. Factional part Sam plus fractional part Chester equals 1 job

$$\frac{1}{8}x + \frac{1}{14}x = 1 \quad \text{The LCD} = (8)(14) = 56$$

$$(56) * \frac{1}{8}x + (56) * \frac{1}{14}x = (56) * 1$$

$$7x + 4x = 56; \quad 11x = 56$$

$$x = \frac{56}{11} \quad = 5\frac{1}{11} \text{ hours to complete the job of spraying.} \quad \text{Answer}$$

63t miles

Car A

24 Miles
Between
Car A and
Car B

55t miles

Distance =Rate*Time

Rate*Time=Distance Traveled by both cars

	Rate*	Time	=Distance
Car A	63 mph	t	63t
Car B	55 mph	t	55t

Car B

Car A
63mph
constant
speed

Car B
55mph
constant
speed

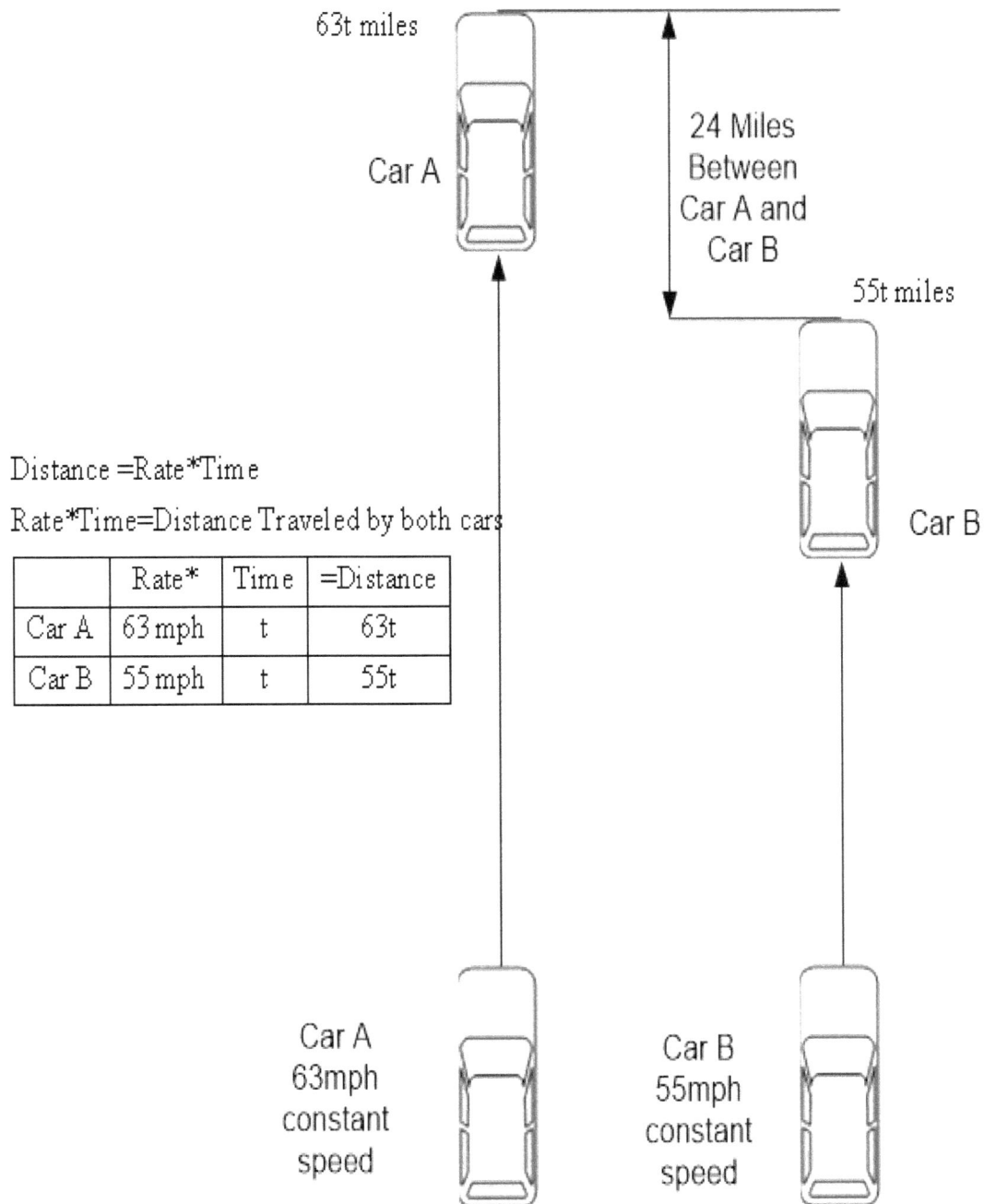

Figure 18.2: Graphical Solution for two cars

18.4   Two Cars algebra word problem

18.5   Solving word problems using equations

18.6   Two car problem in words

Two cars leave Baton Rouge LA at the same time and travel North. Car A travels at 63 mph and Car B travels at 55 mph. In how many hours will be 24 miles apart?

18.7   Analyze the problem

These problems are real life problems solved by the Algebra process.

The first thing to notice is that the time is the same in the problem.

The second is that distance = rate x time.

Also, since this problem cannot easily be done in one step, we need to set up a 2 x 3 matrix of everything we know.

Distance in this case will not initially be solve initially, but we can express distance in Algebraic terms. i.e. 63t and 55t.

	Rate *	Time	= Distance
Car A	63 mph	T	63T
Car B	55 mph	T	55T

Note mph = miles per hour or $\frac{miles}{hour}$

The units in this equation:

Rate: mph in miles

Time: time in hours

Distance = Time in miles; D=rt or rt=d

Miles per hour $= \frac{Miles}{Hour}$ or mph

mph * Hours = Miles

$$\frac{\text{Miles}}{\text{Hour}} * \text{Hours} = \text{Miles}$$

18.8   Formula and Solution

Describe the problem in words

Distance Car A minus (-) Distance Car B= 24 Miles

Translate the words description into an algebraic equation

Use the formula distance = rate*time d=rt

63T-55T=24 Miles

Solve for T hours

8T=24

T=3 hours

Check :

63 ● 3 = 189 miles

55 ● 3 = 165 miles

189 miles - 165 miles = 24 miles

It checks

18.9   Thrown ball from a roof top

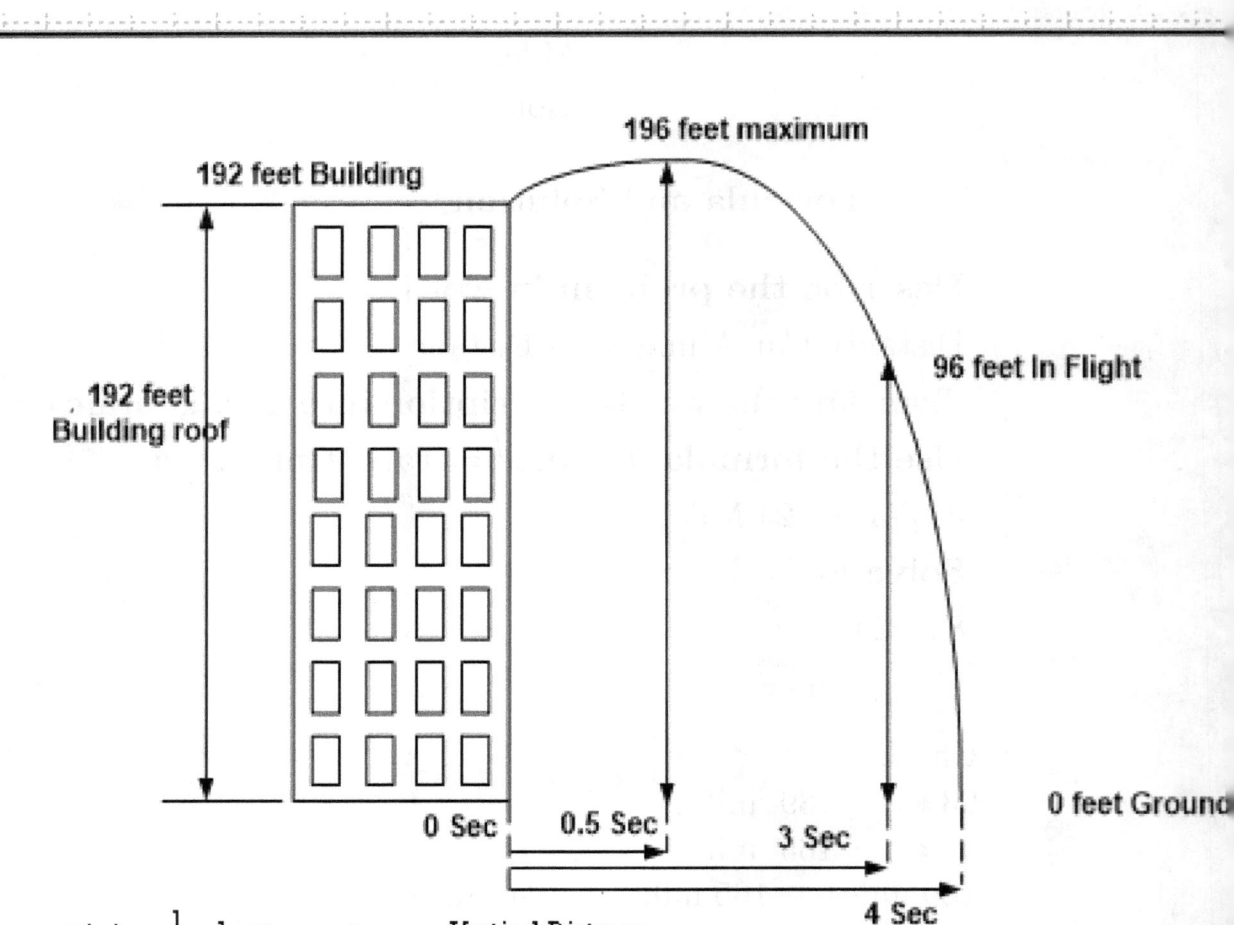

$$y = f(x) = -\frac{1}{2}gt^2 + V_{y0}t + x\,\text{feet}$$

$$f(x) = -16\,t^2 + 16t + 192$$

$$f(x) = -16\,t^2 + 16t + 192$$

Time in seconds	Distance in Feet above ground
0.0	192.0
0.5	196.0
1.0	192.0
1.5	180.0
2.0	160.0
2.5	132.0
3.0	96.0
3.5	52.0
4.0	0

Chapter 19

Calculate the time it takes to hit the ground

A ball is thrown from a roof top with an initial speed of 16 feet/second from the rooftop at 192 feet.

The following equation derivation (from physics) is found on on page 161

$$s(t) = -16t^2 + 16t + 192$$

$s(t)$ is the height dependent on the time (t)

At ground level $s(t) = 0$

$$0 = -16t^2 + 16t + 192$$

solve by dividing both sided by -4

$$4t^2 - 4t - 48 = 0$$

$$(4t + 12)(t - 4) = 0$$

$4t + 12 = 0$

$4t = -12$

$t = -4$ reject a negative time solution

$$(t - 4) = 0$$
$$t = 4 \quad \text{seconds for the ball to hit the ground}$$

19.1 Calculate the height of the vertex

$$x_{vertex} = \left(\frac{-b}{2a}\right)$$

$$\left(\frac{-b}{2a}\right)$$

$$a = -16, b = 16$$

$$x_{vertex} = \left(\frac{-b}{2a}\right) = \frac{(-16)}{2(16)} = \frac{1}{2}$$

$$s(t) = -16t^2 + 16t + 192$$

$$s(t) = y = -16\left(\frac{1}{2}\right)^2 + 16\left(\frac{1}{2}\right) + 192$$

$$y = -4 + 8 + 192 = 196 \text{ feet above ground}$$

The coordinates set of the apex is $\left(\frac{1}{2}, 196\right)$

19.2   Calculate the time it takes to reach 96 feet

A ball is thrown from a roof top with an initial speed of 16 feet/second from the rooftop at 192 feet.

The equation from physics is:

$$s\left(t\right) = -16t^2 + 16t + 192$$

$$s\left(t\right) = 96 \text{ feet}$$

$$96 = -16t^2 + 16t + 192$$

$$0 = -16t^2 + 16t + 192 - 96$$

$$0 = -16t^2 + 16t + 96$$

divide both sides of the equation by (-4)

$$4x^2 - 4x - 24 = 0$$

$$(4x + 8)\left(x - 3\right) = 0$$

$$4x + 8 = 0$$

$$x = -2 \text{ reject solution}$$

Time is always a positive value

$$(x - 3) = 0$$

x = 3 seconds to reach 96 feet

19.3 Pythagorean-theorem-diagram

Pythagorean Theorem
In a right triangle the addition of
the squares "a" and "b"equals
the square
of the hypotenuse "c"
i.e. c^2 = a^2 + b^2

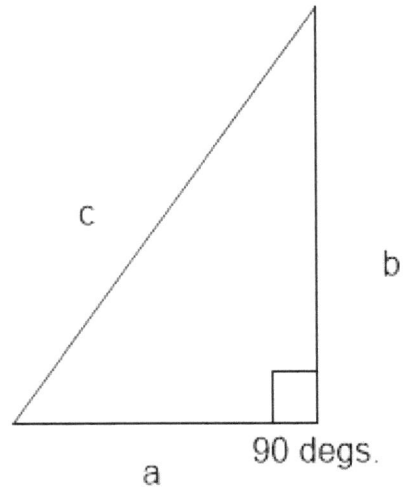

c

b

90 degs.

a

5^2=3^3+2^2
25=9+16
25=25

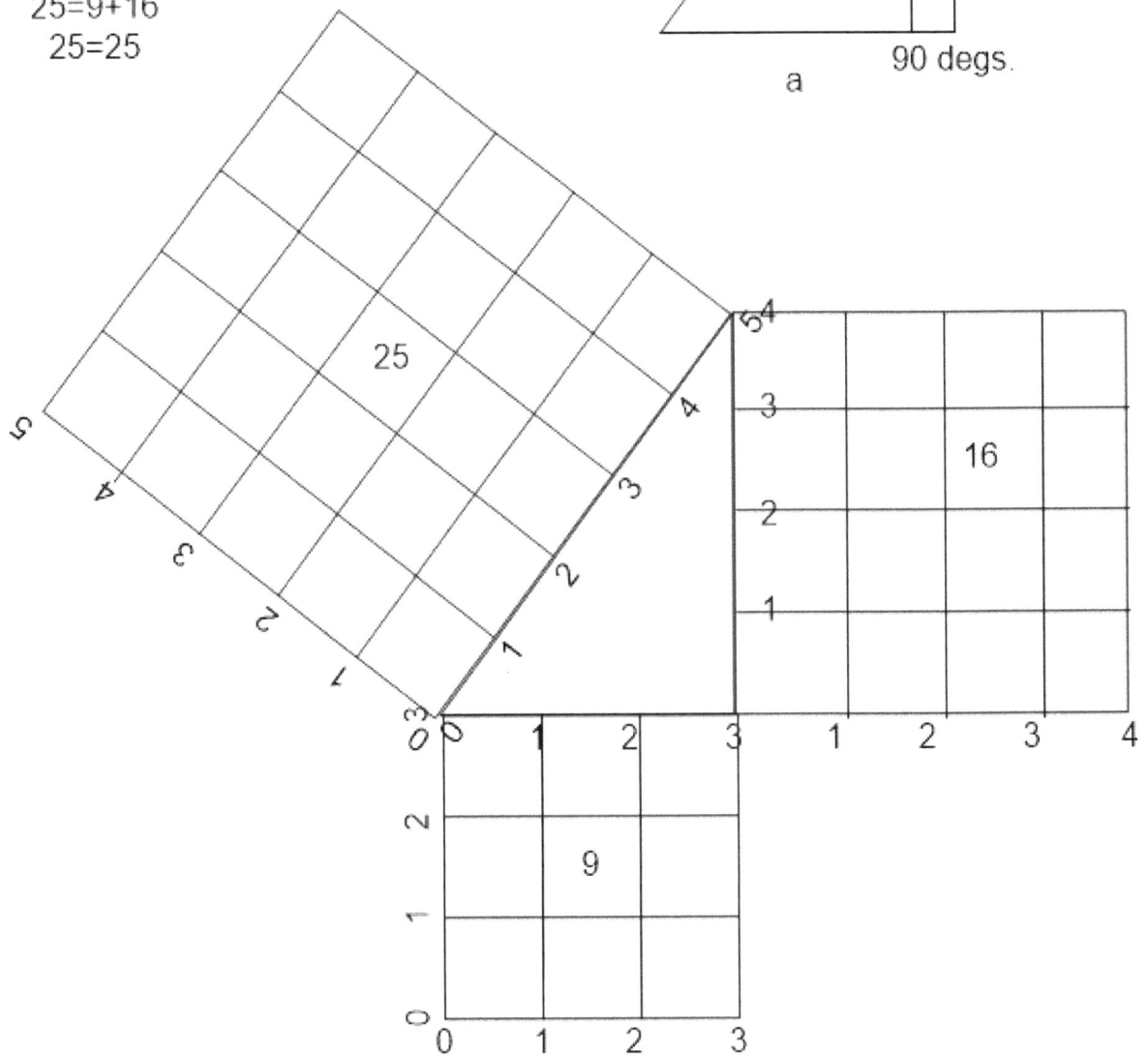

25

16

9

19.4   Pythagorean-theorem-diagra-explanation

A side such as a squared is represented by an area of 4 evenly sized blocks measuring 16 blocks.

A side such as b squared is represented by an area of 3 evenly sized blocks measuring 9 blocks.

A side such as c squared is represented by an area of 5 evenly sized blocks measuring 25 blocks. as shown in the diagram.

Pythagorean Theorem

In a right triangle the addition of the squares "a" and "b" equals the square of the hypotenuse "c"

$$c^2 = a^2 + b^2$$

$$5^2 = 3^2 + 4^2$$

$$25 = 9 + 16$$

$$25 = 25$$

It proves out.

Note: the Pythagorean works for all right triangles.
A right triangle has 90^0 as its largest angle.
Refer to the following diagram.

19.5 Ladder problem description

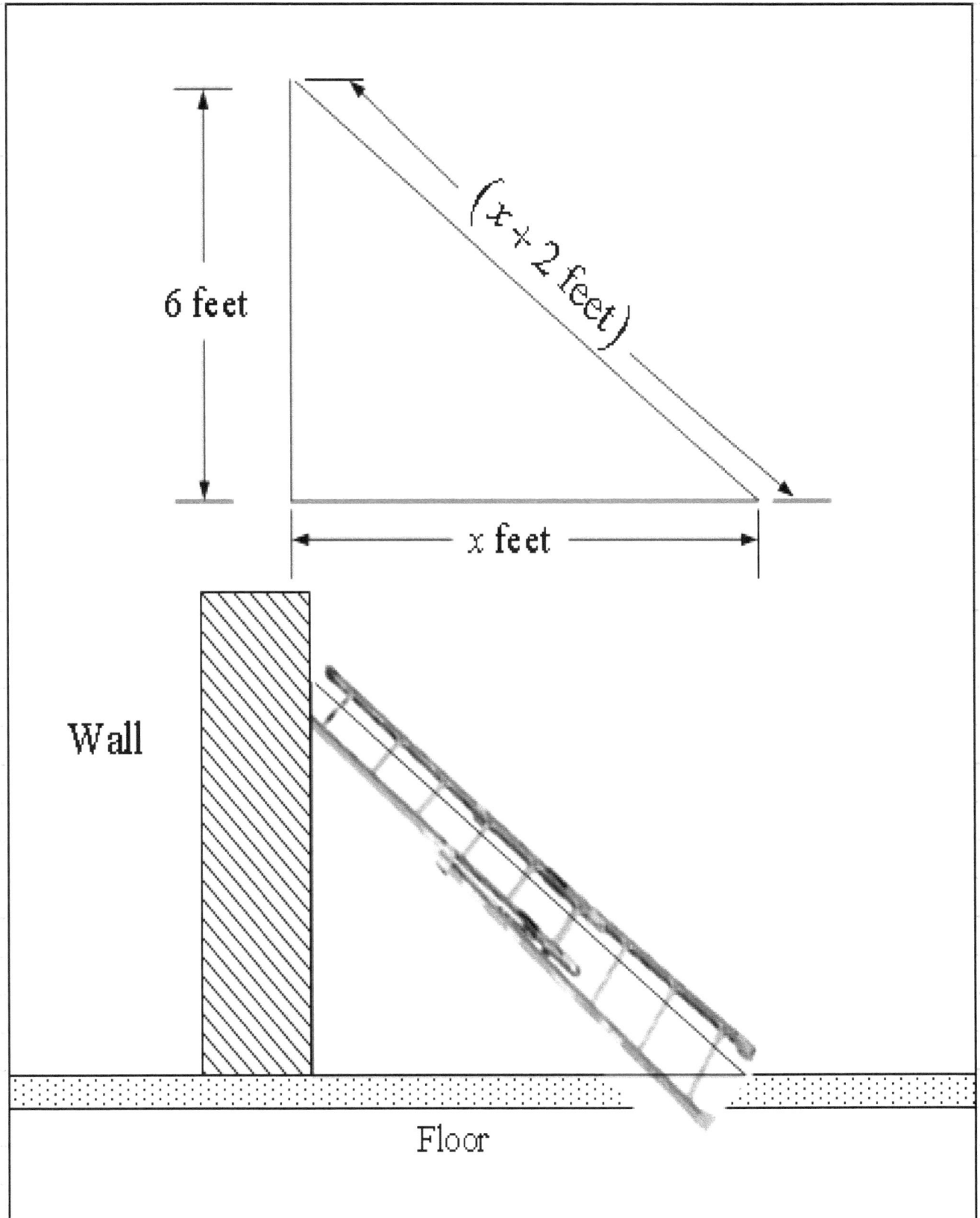

19.6   State the problem

- A ladder is placed on a wall.
- The ladder rests on the wall 6 feet above the floor.
- The bottom of ladder is x feet from the wall equals x
- The ladder is 2 feet longer than the distance from the wall. $x+2$

19.7   Analyze the problem

- Let a = x equal to the short leg of the triangle. The ladder makes a right-angle triangle with the wall.
- Let b = 6 feet the height on the wall where the ladder rests on the wall.
- Let c = (x+2) the length of the ladder.

19.8   Ladder problem solution

We know from the Pythagorean formula that the

hypotenuse squared equals

the sum of the squares of the other two sides.

So we substitute the values of the triangle formed by the ladder and the wall

This is good as far as it goes but what is asked for is the length

of the ladder which equals (x+2)

Therefore, the ladder $= (x + 2) = (8 + 2) = 10$ feet

The ladder is 8 feet from the wall and the ladder is 10 feet long.

$$x^2 + 6^2 = (x + 2)^2$$
$$x^2 + 36 = x^2 + 4x + 4$$
$$x^2 + 4x + 4 = x^2 + 36$$
$$4x + 4 = 36$$
$$4x = 32$$
$$x = 8$$
$$\text{and}$$
$$x + 2$$
$$x + 2 = 10$$

Chapter 20

Types of quadratic equation solutions

20.1   Example of a real solution

The quadratic formula uses the coefficients of the quadratic-equation to solve for x.

$$x = \frac{-b \pm \sqrt{b^2 - 4ac}}{2a}$$

General Format

$$ax^2 + bx + c$$

Typical Quadratic equation

$$x^2 + 4x + 4$$

Factored Expression

$$x^2 + 4x + 4 = (x + 2)(x + 2)$$

20.2   Use factor method

Where If factorable, factor into 2 binomials

If not factorable use the quadratic formula

Note: The Quadratic formula works in both cases.

20.3   Another Quadratic equation

$$x^2 + 2x - 15 = 0$$

$$(x + 3)(x - 5) = 0$$

$$(x + 3) = 0$$
$$x = -3$$

$$(x - 5) = 0$$
$$x = 5$$

This graph can be found of page 118

20.4 Solve the previous quadratic equation

Solve the same quadratic equation
using the quadratic formula.
We get the same solution $x = -3$ and $x = 5$

Coefficants and constant for setting up binomial formula

$$x^2 + 2x - 15 = 0$$
$$a = 1, \ b = 2, \ c = -15$$

Using the quadratic formula:

$$x = \frac{-b \pm \sqrt{b^2 - 4ac}}{2a}$$

$$x = \frac{-2 \pm \sqrt{2^2 - 4\,(1)\,(-15)}}{2\,(1)}$$

$$x = \frac{-2 \pm \sqrt{64}}{2}$$

$$x = \frac{-2 \pm 8}{2}; \ \text{Therefore} \quad x = \frac{-2 + 8}{2} \quad x = \frac{-2 - 8}{2}$$

$$x = \frac{6}{2} = 3 \quad x = \frac{-10}{2} = -5 \text{In summary } x = 3, -5$$

This plot or graph can be fount on page118 We can verify the solution
by plotting the equation: We name the

$$\sqrt{b^2 - 4ac}$$

the determinate of the equation.
The result of solving the determinate yields one of 3 possible solution
real, imaginary or no real solution (in the case of the square root being
negative.)

20.5   Imaginary number system

An imaginary number system consists of x and iy.

Where $i = \sqrt{-1}$

Imaginary numbers are plotted on the y -axis only.

A typical coordinate would be $\{x, y\} = \{5x, 3iy\}$

20.6   Example of an imaginary solution

$$3x^2 - 5x + 4 \text{ where a} = 3, b = -5, \ c = 4$$

$$x = \frac{-b \pm \sqrt{b^2 - 4ac}}{2a}$$

$$x = \frac{-(-5) \pm \sqrt{(-5)^2 - 4(3)(4)}}{2(3)}$$

$$x = \frac{5 \pm \sqrt{-23}}{6} = x = \frac{5 \pm 4.796\, i}{6} \quad \text{where } i \quad \sqrt{-1}$$

therefore, the roots are imaginary and do not intersect with the real number line.

This graph can be found on page **??**

20.7 Plotting a quadratic with and imaginary solution

Note: Imaginary solutions does not mean the graph of the equation does not exist. It does however mean that the parabola does not cross the real number line (the +-x-axis) hence the solution itself does not exist.

20.8 Explanation of the determinate

$$x = \frac{-b \pm \sqrt{b^2 - 4ac}}{2a}$$

The determinate is the radical part of the quadratic formula and the result determines the type of solution.
The determinate is

$$\sqrt{b^2 - 4ac}$$

- Real solutions if Determinate is greater than or equal to zero.

- If equation is equal zero then the apex (maximum or minimum) intersects with the x axis and there are 2 solutions that are the same i.e. one solution

- If the determinate is less than zero then there are imaginary solutions and the entire parabola is either above or below the x-axis.

- Note: parabolas with imaginary solutions are still able to be represented on a graph. (Cartesian coordinate graph)

- The quadratic formula as a whole solves for

- x depending of the coefficients

- of the equation.

- Calculating the expression,

- $\left(\frac{-b}{2a}\right)$

- will solve the apex of the quadratic.

- Calculating the $\sqrt{b^2 - 4ac}$ is used to determine whether the root is real, zero, or imaginary.

20.9 Examples of quadratic equations

Examples of quadratic equations, are throwing a ball up in the air with an upward motion and calculating the results.

The problem stated, is throwing in the x-direction (upward).

Chapter 21

Real Solution

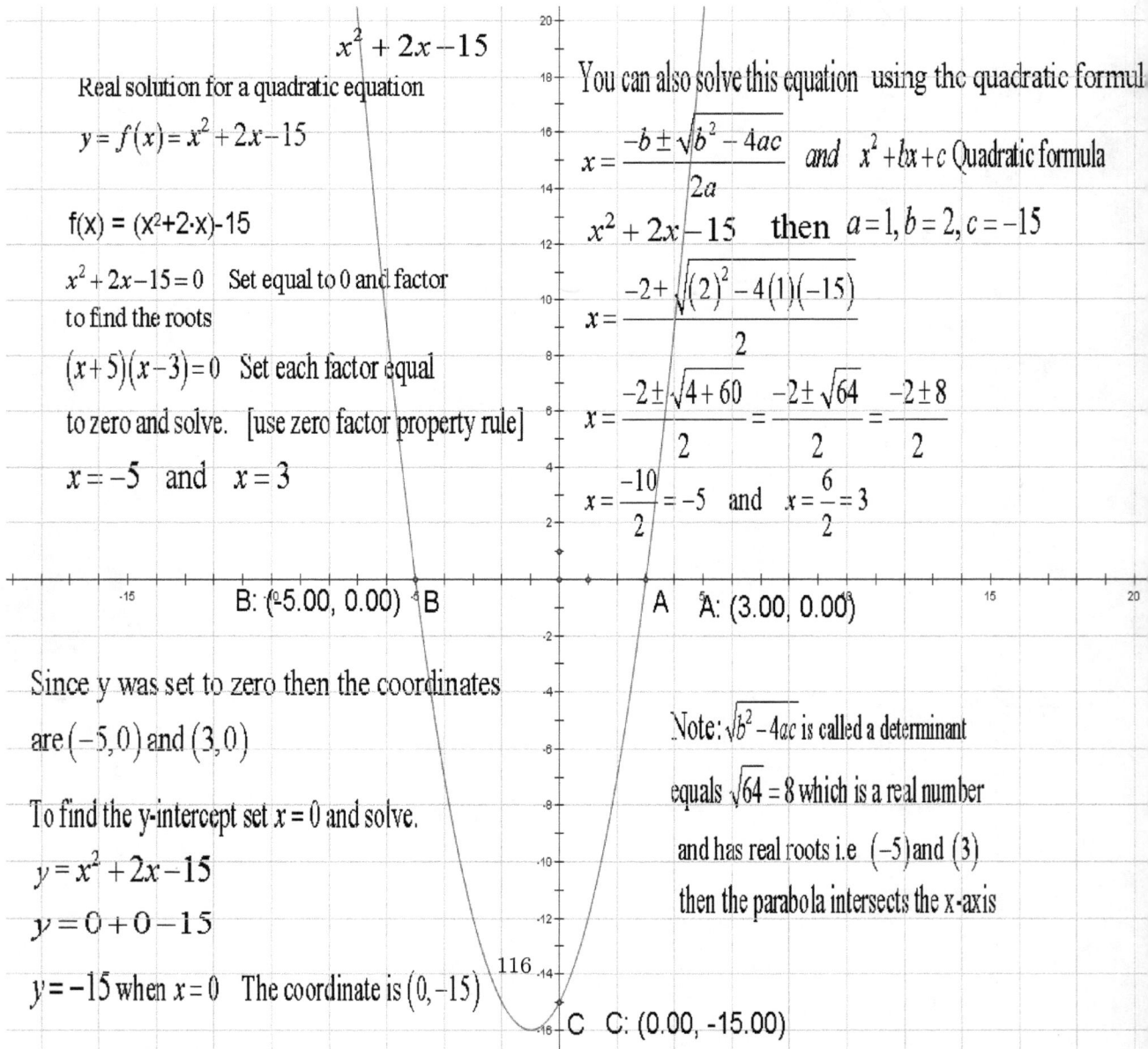

$$x^2 + 2x - 15$$

Real solution for a quadratic equation

$$y = f(x) = x^2 + 2x - 15$$

f(x) = (x²+2·x)-15

$x^2 + 2x - 15 = 0$ Set equal to 0 and factor to find the roots

$(x+5)(x-3) = 0$ Set each factor equal to zero and solve.   [use zero factor property rule]

$x = -5$ and $x = 3$

You can also solve this equation using the quadratic formula

$$x = \frac{-b \pm \sqrt{b^2 - 4ac}}{2a} \quad and \quad x^2 + bx + c \text{ Quadratic formula}$$

$x^2 + 2x - 15$ then $a = 1, b = 2, c = -15$

$$x = \frac{-2 + \sqrt{(2)^2 - 4(1)(-15)}}{2}$$

$$x = \frac{-2 \pm \sqrt{4 + 60}}{2} = \frac{-2 \pm \sqrt{64}}{2} = \frac{-2 \pm 8}{2}$$

$$x = \frac{-10}{2} = -5 \quad and \quad x = \frac{6}{2} = 3$$

B: (-5.00, 0.00) B   A   A: (3.00, 0.00)

Since y was set to zero then the coordinates are $(-5, 0)$ and $(3, 0)$

To find the y-intercept set $x = 0$ and solve.

$y = x^2 + 2x - 15$

$y = 0 + 0 - 15$

$y = -15$ when $x = 0$ The coordinate is $(0, -15)$

Note: $\sqrt{b^2 - 4ac}$ is called a determinant

equals $\sqrt{64} = 8$ which is a real number

and has real roots i.e (-5) and (3)

then the parabola intersects the x-axis

116

C   C: (0.00, -15.00)

Figure 21.1: Real Solution

21.1   Quadratic equation Imaginary solution

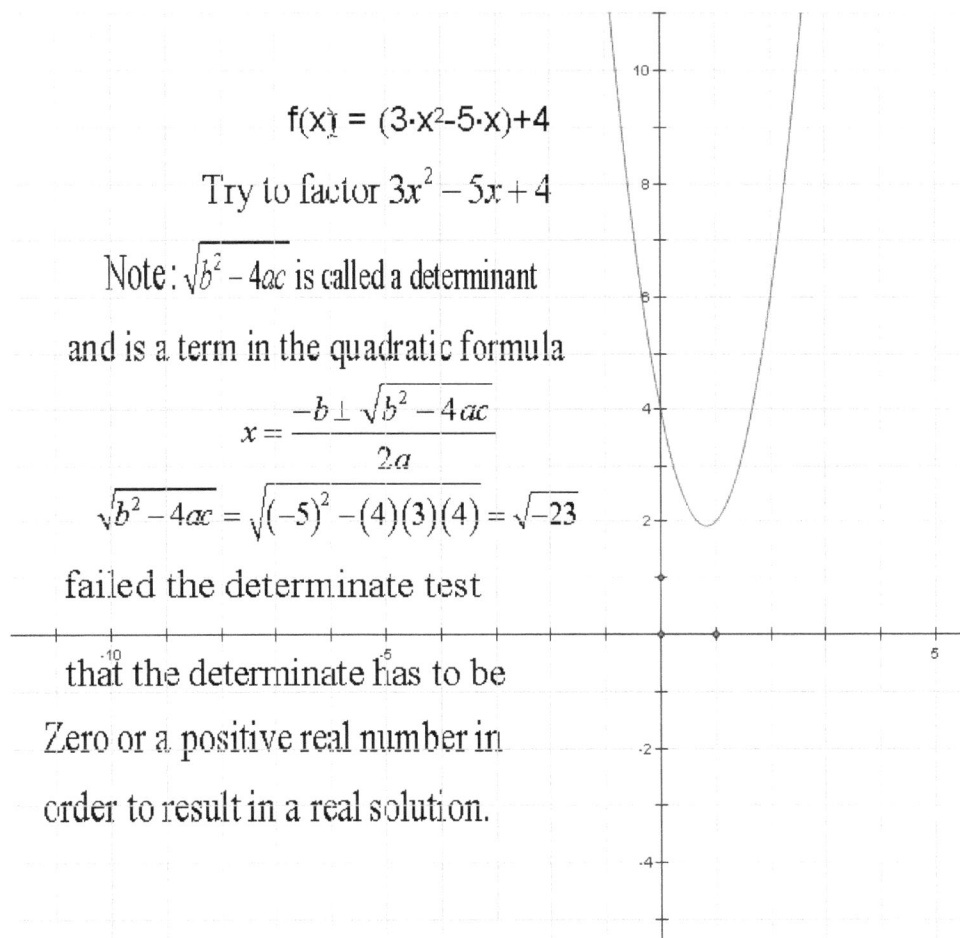

$f(x) = (3 \cdot x^2 - 5 \cdot x) + 4$

Try to factor $3x^2 - 5x + 4$

Note: $\sqrt{b^2 - 4ac}$ is called a determinant

and is a term in the quadratic formula

$$x = \frac{-b \perp \sqrt{b^2 - 4ac}}{2a}$$

$$\sqrt{b^2 - 4ac} = \sqrt{(-5)^2 - (4)(3)(4)} = \sqrt{-23}$$

failed the determinate test

that the determinate has to be

Zero or a positive real number in

order to result in a real solution.

Chapter 22

Rules of Fractional Exponents

- The m's and n's are integers exponents.

- Note: The $x's$ are variables, real numbers, or algebraic expressions

- when working with the following rules.

- Rational exponents as fractional exponents

- If a is a real number and n is a positive integer such that the principal n th root os a exists then $a^{\frac{1}{n}} = \sqrt{a}$ by defintion where $\frac{1}{n}$ is the fractional exponent of a.

- Also ,if m is a positive integer with no common factor with n then:

- $a^{\frac{m}{n}} = \left(a^{\frac{1}{n}} \right)^m = \sqrt[n]{a}^m$

- $\sqrt{100} = \pm 10$

- $\sqrt{64} = \pm 8$

Chapter 23

Formal rules of exponents

	Property	Examples
1)	$x^m x^n = x^{m+n}$	$3^3 \bullet 3^5 = 3^{3+5} = 3^8 = 6561$
2)	$\frac{x^m}{x^n} = x^{m-n}$	$\frac{x^8}{x^3} = x^{8-3} = x^5$
3)	$x^{-n} = \frac{1}{x^n}$	$x^{-4} = \frac{1}{x^4}$
4)	$x^0 = 1$	$\left(x^2 + 2\right)^0 = 1$
5)	$(xy)^n = x^n y^n$	$(3x)^3 = 3^3 x^3 = 27x^3$
6)	$\left(x^m\right)^n = x^{mn}$	$\left(x^4\right)^{-5} = x^{4(-5)} = x^{-20} = \frac{1}{x^{20}}$
7)	$\left(\frac{x}{y}\right)^m = \frac{x^m}{y^m}$	$\left(\frac{3}{x}\right)^2 = \frac{3^2}{x^2} = \frac{9}{x^2}$
8)	$\left|x^2\right| = \left|x\right|^2 = x^2$	$\left|\left(-2^2\right)\right| = \left|-2\right|^2 = (2)^2 = 4$

Chapter 24

Square root characteristics

- The set of square coordinates are (zero, zero), or (positive, positive).

- The square root of a number is always a real number

- therefore, a negative root will not yield a real number.

- Non-perfect square roots will yield irrational value numbers.

- Example of two non-perfect numbers.

- Note: None of the set of coordinates are in

- the negative x section of the graph.

- You cannot have the square root of a negative number and have a real result.

- The result will be imaginary

- Perfect positive square roots will yield Whole numbers.

24.1   Table of first 5 perfect square roots

x	$y = (x) = \sqrt{x}$	$(x, y) \, or \, (f(x))$
0	$y = f(0) = \sqrt{0} = 0$	$(0,0)$
1	$y = f(1) = \sqrt{1} = 1$	$(1,1)$
4	$y = f(4) = \sqrt{4} = 2$	$(4,4)$
9	$y = f(9) = \sqrt{9} = 3$	$(9,3)$
16	$y = f(16) = \sqrt{16} = 4$	$(16,4)$

Figure 24.1: First five perfect square roots

24.2   Graph of first five perfect square roots

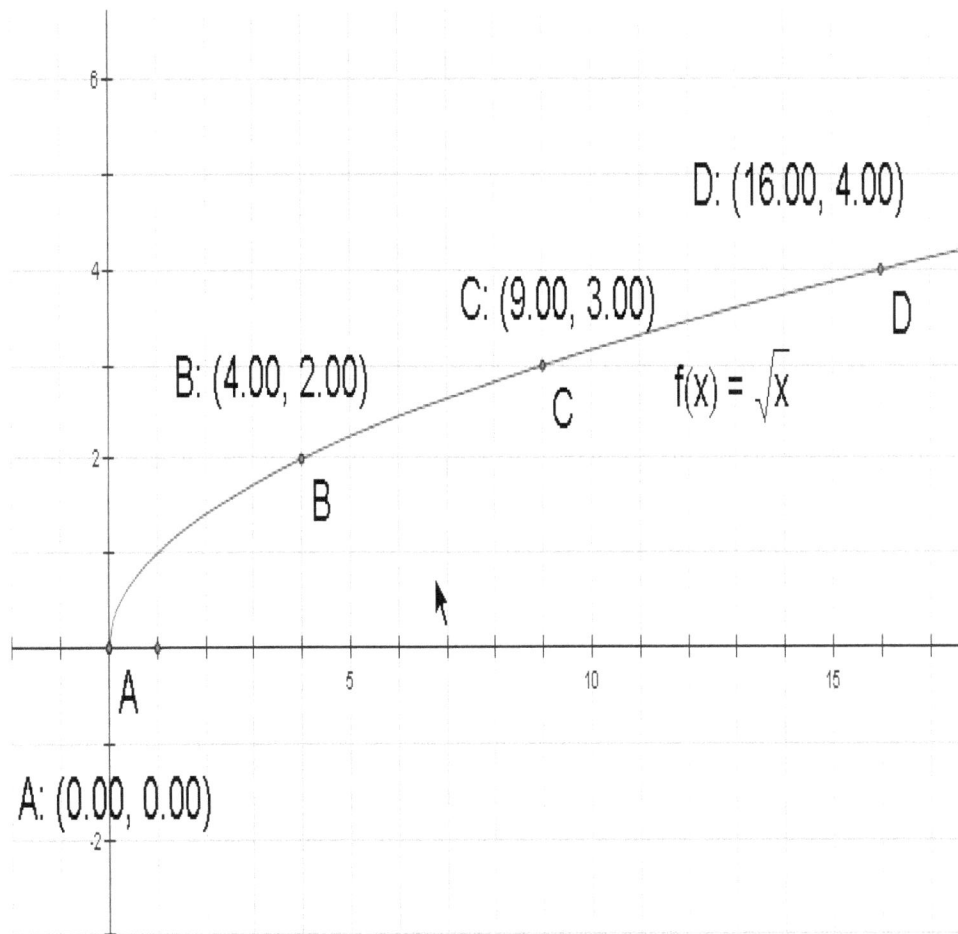

Figure 24.2: Graph perfect square roots

Chapter 25

Cube root characteristics

- The cube root can be in the negative quadrant and the positive quadrant or zero.

- The set of cubic coordinates are (negative, negative), or (zero, zero), or (positive, positive).

- The cube root of a number is always a real number,therefore a negative root as well as a zero root or a positive root will yield a real number.

- Non-perfect cube roots will yield an irrational value number.

- Perfect cube roots will yield a Whole number

- The cube root of x can be expressed as a fractional exponent.

- $\sqrt[3]{x} = x^{\frac{1}{3}}$

- Therefore, we graph $x^{\frac{1}{3}}$ as a cube root. See graph.

- $h(x) = x^{\frac{1}{3}}$

25.1   Table of first five perfect cube roots

x	$y = (x) = \sqrt[3]{x}$	$(x,y)\, or\, (f(x))$
-8	$y = f(-8) = \sqrt[3]{-8} - 2$	$(-8,-2)$
-1	$y = f(-1) = \sqrt[3]{-1}$	$(-1,-1)$
0	$y = f(0) = \sqrt{0} = 0$	$(0,0)$
1	$y = f(1) = \sqrt{1} = 1$	$(1,1)$
8	$y = f(8) = \sqrt[3]{8} = 2$	$(8,2)$

Figure 25.1: Table of first five perfect cube roots

25.2 Graph of first five perfect cube roots

Graph of first five perfect cube roots

$y = \sqrt[3]{x}$ y equals the cube root of x

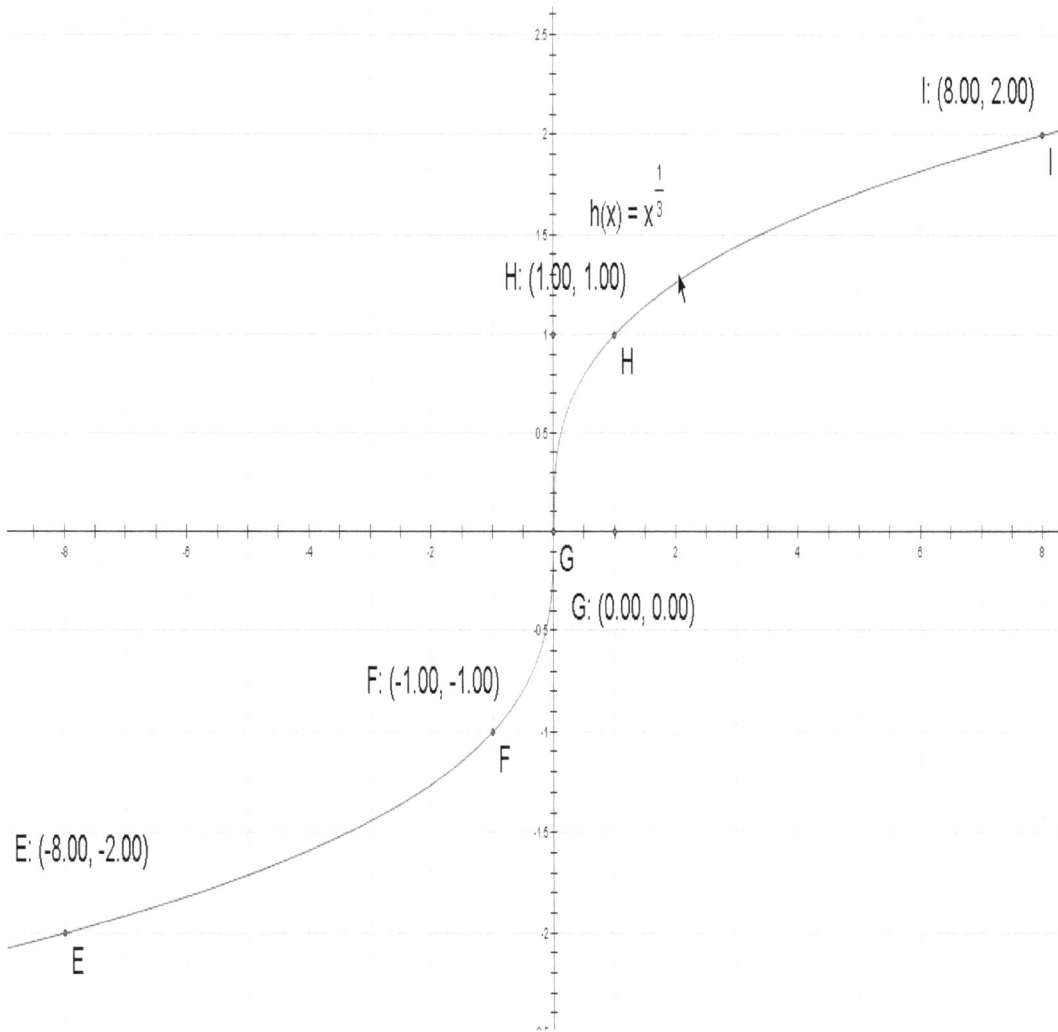

Figure 25.2: First five perfect cube roots

Chapter 26

Rational Expressions and Complex Fractions

There are two ways to solve complex fractions

- Case 1: Use Lowest common denominator separately with the numerator.

- A Case 1 Easier method of solving complex fractions is found on page 137

- Case 2: Use Lowest common denominator of the entire rational expression..

- Case 2: follows

$$\frac{\dfrac{1}{m-1} + \dfrac{2}{m+2}}{\dfrac{2}{m+2} - \dfrac{1}{m-3}}$$

Find the lowest common denominator of all the denominators
  The Denominators are (m-1) and (m-3) and (m+2),(m+2)
Note" For the lowest common denominator (m+2) only gets

counted once since it's highest power (exponent) is one.

- The basic principle of adding rational expressions is that rational expressions can be added or subtracted, when the denominators are the same.

- To do that we need the lowest common denominator.

- The factors are counted the least time that they appear.

- Denominators (m-1) and (m+2) and (m-3)

- are only counted one time.

- The factors are (m-1)(m+2)(m-3) or the lowest common denominator;

- Note" For the lowest common denominator (m+2) only gets counted once.

- If the denominator was for instance (m-3) then we would need to multiply by (m-1)(m+2).

- We need to multiply by the ratio of Missing-factors Ł

$$\frac{missing - factor(s)}{missing - factor(s)}$$

which is

$$\frac{(m-1)(m+2)}{(m-1)(m+2)}$$

or

$$\frac{(m+2)(m-3)}{(m+2)(m-3)}$$

or

$$\frac{(m-1)(m-3)}{(m-1)(m-3)}$$

depending on the denominator which will result in the denominators being the same which is

$$(m-1)(m+2)(m-3)$$

- This is algebraically the same as multiplying by 1 since the same
numerator and denominator equals one (1) one and will not change the value of the fraction when multiplying by the equivalent of one.

Complex Equation to be solved

$$\frac{\frac{1}{m-1} + \frac{2}{m+2}}{\frac{2}{m+2} - \frac{1}{m-3}}$$

Use sets of Missing-factors used to create the **LCD (lowest common denominator)**

The Lowest common denominator in this example is:

$$(m-1)(m+2)(m-3)$$

Note: To take the first example of

$$\frac{(m-1)(m+2)}{(m-1)(m+2)}$$

Note That:

$$\frac{Missing - factors}{Missing - factors} = \frac{(m-1)(m+2)}{(m-1)(m+2)} = 1 = \text{ a unity value}$$

Therefore, multiplying by a unity value (one) does not change the value of an expression although the appearance will change. Also, since this is one of the missing-factors it will create the factor

$$(m - 1)(m + 2)(m - 3)$$

in the denominator of one of the of the sets. Similarly, we also multiply by the

$$\frac{\text{Missing} - \text{factors}}{\text{Missing} - \text{factors}}$$

to create the denominator in the correct factors (lowest common denominator).

Note: we also multiply the missing factor(s) in the numerator creating a new correct expression.

Complex Fraction to be solved

$$\frac{\dfrac{1}{m-1} + \dfrac{2}{m+2}}{\dfrac{2}{m+2} - \dfrac{1}{m-3}}$$

Multiply by

$$\frac{\text{Missing} - \text{factors}}{\text{Missing} - \text{factors}}$$

$$\frac{\dfrac{1}{m-1}\dfrac{(m+2)(m-3)}{(m+2)(m-3)} + \dfrac{2}{m+2}\dfrac{(m-1)(m-3)}{(m-1)(m-3)}}{\dfrac{2}{m+2}\dfrac{(m-1)(m-3)}{(m-1)(m-3)} - \dfrac{1}{m-3}\dfrac{(m-1)(m+2)}{(m-1)(m+2)}}$$

$$\frac{\dfrac{1}{m-1}\dfrac{\left(m^2+2m-3m-6\right)}{(m+2)(m-3)} + \dfrac{2}{m+2}\dfrac{\left(m^2-m-3m+3\right)}{(m-1)(m-3)}}{\dfrac{2}{m+2}\dfrac{\left(m^2-m-3m+3\right)}{(m-1)(m-3)} - \dfrac{1}{m-3}\dfrac{\left(m^2-m+2m-2\right)}{(m-1)(m+2)}}$$

Now that when the denominators are the same, we keep the denominators and add or subtract the terms in the numerator

$$\frac{\dfrac{3m^2-9m}{(m-1)(m+2)(m-3)}}{\dfrac{2}{m+2}\dfrac{\left(2m^2-2m-6m+6\right)}{(m-1)(m-3)} - \dfrac{1}{m-3}\dfrac{\left(m^2-m+2m-2\right)}{(m-1)(m+2)}}$$

Now do the same for the denominator in the numerator, and the denominator in the denominator.

$$\frac{\dfrac{3m^2-9m}{(m-1)(m+2)(m-3)}}{\dfrac{2m^2-2m-6m+6-m^2-m+2}{(m+2)(m-1)(m-3)}} = \frac{\dfrac{3m^2-9m}{(m-1)(m+2)(m-3)}}{\dfrac{m^2-9m+8}{(m+2)(m-1)(m-3)}}$$

To resolve the large fraction bar, invert the denominator and multiply times the numerator

$$\frac{3m^2-9m}{(m-1)(m+2)(m-3)} * \frac{(m+2)(m-1)(m-3)}{m^2-9m+8} =$$

$$\frac{3m^2-9m}{(m-1)(m+2)(m-3)} * \frac{(m+2)(m-1)(m-3)}{m^2-9m+8} =$$

$$\frac{3m^2-9m}{m^2-9m+8}$$

$$\frac{3m^2-9m}{(m-1)(m-8)}\text{ans}$$

Chapter 27

Simpler method of solving complex fractions

The reason the solution is simpler, is that we only us
two binary expressions at one time.
Before the last step we need to rationalize the denominator
with the third missing factor. This is diffenent for each expression.

A simpler method of solving a complex fraction

is to find the lowest common denominator

separately for the numerator for the

fraction and the denominator of the fraction.

Solve this example.

$$\frac{\dfrac{1}{m-1} + \dfrac{2}{m+2}}{\dfrac{2}{m+2} - \dfrac{1}{m-3}}$$

The $\dfrac{missing-factor(s)}{missing-factor(s)}$ are $\dfrac{m+2}{m+2}$ or $\dfrac{m-1}{m-1}$

for the numerator. and

The $\dfrac{missing-factor(s)}{missing-factor(s)}$ are $\dfrac{m+2}{m+2}$ or $\dfrac{m-3}{m-3}$

for the denominator.
Multiply by

$$\frac{missing\ factor(s)}{missing\ factor(s)}$$

$$\frac{\dfrac{1}{m-1}\dfrac{(m+2)}{(m+2)} + \dfrac{2}{m+2}\dfrac{(m-1)}{(m-1)}}{\dfrac{2}{m+2}\dfrac{(m-3)}{(m-3)} - \dfrac{1}{m-3}\dfrac{(m+2)}{(m+2)}}$$

Collect terms in the numerator and denominator fractions Rationalize the denominater by multiply by the $\frac{(missing\,factor)}{(missing\,factor)}$ Note the missing factor is different for each fraction. Now add the numerators and keep the denominator for each fraction i.e (rational factor) where a rational expression is algebraic fraction.

$$\frac{\dfrac{m+2}{m-1(m+2)} + \dfrac{2m-2}{m+2)(m-1)}}{\dfrac{2m-6}{(m+2(m-3)} - \dfrac{(m+2)}{(m+2)(m-3)}}$$

$$\frac{\dfrac{m+2+2m-2}{(m+2)(m-1)}}{\dfrac{(2m-6)-(m+2)}{(m+2)(m-3)}} \quad = \quad \frac{\dfrac{3m}{(m+2)(m-1)}}{\dfrac{m-8}{(m+2)(m-3)}}$$

Invert the denominator and multiply by 3rd missing factor $\frac{m-3}{m-3}$ or $\frac{m-1}{m-1}$

$$\frac{3m}{(m+2)(m-1)} * \frac{(m-3)}{(m-3)} = \frac{(3m^2-9m)}{(m-1)(m-3)(m+2)}$$

$$\frac{m-8}{(m+2)(m-3)} * \frac{(m-1)}{(m-1)} = \frac{(m-8)(m-1)}{(m-1)(m-3)(m+2)}$$

Denominator inverted

$$\frac{(m-1)(m+2)(m-3)}{(m-8)(m-1)}$$

Numerator times inverted denominator

$$\frac{(3m^2-9m)}{(m-1)(m-3)(m+2)} * \frac{(m-1)(m+2)(m-3)}{(m-8)(m-1)}$$

Answer $\frac{(3m^2-9m)}{(m-8)(m-1)}$

Chapter 28

Proof for Complex-Fractions

By picking a constant and substituting the constant into both the problem and the solution and checking for a match.

Problem

$$\frac{\frac{1}{m-1} + \frac{2}{m+2}}{\frac{2}{m+2} - \frac{1}{m-3}}$$

Solution

$$\frac{3m^2 - 9m}{(m-1)(m-8)} \text{ans}$$

Prove that the solution is the same as the problem by substitution of a variable for x where x not equal to 1 or 3 avoiding a zero term.

Problem = Solution ?

Does

$$\frac{\frac{1}{m-1} + \frac{2}{m+2}}{\frac{2}{m+2} - \frac{1}{m-3}}$$

equal

$$\frac{3m^2 - 9m}{(m-1)(m-8)}$$

Problem

$$\frac{\dfrac{1}{2-1} + \dfrac{2}{2+2}}{\dfrac{2}{2+2} - \dfrac{1}{2-3}}$$

$$\frac{1 + 1/2}{1/2 + 1} = 1$$

Solution

$$\frac{-6}{(1)(-6)} = 1$$

Problem = Solution

Yes: Substituting 2 for x in the problem yields 1   Yes Substituting 2 for x in the solution yields 1

Therefore since 1=1 The solution is correct Q E D Latin for "quod erat demonstrandum", meaning "which was to be demonstrated".

Chapter 29

Definition of a Radical

Radical $= \sqrt[n]{a}$

Where n equals the index.

Where $\sqrt{}$ equals the radical sign and

Where $\sqrt[n]{a}$ equals the entire radical

Where a is the radicand under the radical sign.

29.1 Examples of radicals

Radicals are expressions under a root sign.

$\sqrt{x}$ or $\sqrt{x+2}$ are two different examples.

$\sqrt[2]{x}$ is expressed as $\sqrt{x}$ where the second root is implied

Higher roots than 2 such as 3, in $\sqrt[3]{x}$ are always specified.

Chapter 30

Adding-and-subtracting-radicals

30.1 Add or subtract Radicals

If a radical expression contains

contains sums or differences of Radicals

that are the same under the root sign,

add the coefficients and keep the root.

I.E. $3\sqrt{2} + 4\sqrt{2} = 7\sqrt{2}$

30.2 Resolving Radicals

After simplifying by extracting,

the perfect squares

the expressions both

have the same

remaining radical $\sqrt{2}$

Simplify $5 \bullet (\sqrt{8} - \sqrt{32})$

$5 \bullet (\sqrt{8} - \sqrt{32}) = 5(\sqrt{(4)(2)} - \sqrt{(16)(2)})$

$5(2\sqrt{2} - 4\sqrt{2}) = -10\sqrt{2})$

Chapter 31

Rationalizing-the-denominator

The radical expression $\sqrt{\frac{3}{2}} = \frac{\sqrt{3}}{\sqrt{2}}$ is rationalized
by multiplying the radical expression by the unity-factor
$\frac{\sqrt{2}}{\sqrt{2}}$.

$$\frac{\sqrt{3}}{\sqrt{2}} * \frac{\sqrt{2}}{\sqrt{2}} = \frac{\sqrt{2}\sqrt{3}}{\sqrt{2}\sqrt{2}} = \frac{\sqrt{6}}{2},$$

which satisfies the requirement that in a solved radical expression, the solved denominator cannot contain a radical.

31.1 Exp.2 Rationalizing Radicals denominators.

Using the equalization principle in the previous example:

$$\frac{\sqrt{2}}{\sqrt{2}} = 1$$

is a unity factor, and multiplying by a unity-factor does not change the value of the expression. Multiplying by a unity-factor produces an equivalent form of the expression. Therefore,

$$\sqrt{\frac{3}{2}} = \frac{\sqrt{3}}{\sqrt{2}} = \frac{\sqrt{3}}{\sqrt{2}} * \frac{\sqrt{2}}{\sqrt{2}} = \frac{\sqrt{6}}{2}$$

is a valid process for rationalizing the denominator.

31.2   An example of multiplying radicals

$$\frac{\sqrt{6x^2y}}{3}$$

Separate the x^2 into a separate radical factor

Resolve $\sqrt{x^2} = x$

$$\frac{\sqrt{x^2}\sqrt{6y}}{3} = \frac{x\sqrt{6y}}{3} \quad \text{Answer}$$

31.3   Rationalizing denominators with a cubic root.

$$\frac{\sqrt[3]{3}}{\sqrt[3]{2}} = \frac{\sqrt[3]{3}}{\sqrt[3]{2}} * \frac{\sqrt[3]{4}}{\sqrt[3]{4}} = \frac{\sqrt[3]{3*4}}{\sqrt[3]{2*4}} = \frac{\sqrt[3]{3*4}}{\sqrt[3]{8}} = \frac{\sqrt[3]{12}}{\sqrt[3]{8}} = \frac{\sqrt[3]{12}}{2}$$

Multiply the fraction by a unity factor that will make the denominator a perfect cube

$$\frac{\sqrt[3]{4}}{\sqrt[3]{4}}$$

and take the cube root of the denominator to simplify.

We can also express the result as

$$\frac{\sqrt[3]{12}}{2} = \frac{1}{2}\sqrt[3]{12}$$

31.4 Multiply an integer times a radical expression

Simplify $5(\sqrt{8} - \sqrt{32})$ First factor out $\sqrt{4}$ the perfect square of $5(\sqrt{8} - \sqrt{32})$ in the first expressions and remaining in both expressions producing a like radical in both terms

$$5\left(\sqrt{(4)(2)} - \sqrt{(16)(2)}\right) = 5\left(2\sqrt{2} - 4\sqrt{2}\right) = 10\sqrt{2} - 20\sqrt{2} = -10\sqrt{2}$$

31.5 Rationalizing with a radical in the denominator

$$\frac{5}{\sqrt{3}} = \frac{5}{\sqrt{3}} * \frac{\sqrt{3}}{\sqrt{3}} = \frac{5\sqrt{3}}{3} = \frac{5}{3}\sqrt{3}$$

$$\frac{\sqrt{3}}{\sqrt{2}} = \left(\frac{\sqrt{3}}{\sqrt{2}} * \frac{\sqrt{2}}{\sqrt{2}}\right) = \frac{\sqrt{6}}{2} = \frac{1}{2}\sqrt{6}$$

- Principals: Multiplying by the $\frac{\sqrt{2}}{\sqrt{2}}$ is algebraically the same as multiplying by one, since $\frac{\sqrt{2}}{\sqrt{2}}$ is a unity factor,

- Multiplying by

$$\frac{Numerator1}{\sqrt{2}} * \frac{Numerator2}{\sqrt{2}} = \frac{New\ Numerator\ (1*2)}{2}$$

resulting with 2 in the denominator. Since $\left(\sqrt{2}\right)^2 = \left(2^{\frac{1}{2}}\right)^2 = (2)^1 = 2$ in the denominator.

- Multiplying by a unity factor does not change the value of the expression and creates an equivalent form of the expression which can be now solved.

- In a solved radical expression, the denominator cannot contain a radical.

31.6 Exp.2 Rationalizing Radicals denominators.

31.7 Rationalizing denominators with a cubic root.

$$\frac{\sqrt[3]{3}}{\sqrt[3]{2}} = \frac{\sqrt[3]{3}}{\sqrt[3]{2}} * \frac{\sqrt[3]{2^2}}{\sqrt[3]{2^2}} = \frac{\sqrt[3]{3 * 2^2}}{\sqrt[3]{2 * 2^2}} = \frac{\sqrt[3]{3 * 2^2}}{\sqrt[3]{2^3}} = \frac{\sqrt[3]{12}}{2} = \frac{1}{2}\sqrt[3]{12}$$

Multiply the fraction by a unity factor that will make the denominator a perfect cube

$$\frac{\sqrt[3]{2^2}}{\sqrt[3]{2^2}}$$

and take the cube root of the denominator to simplify.

We can express the result as

$$\frac{\sqrt[3]{12}}{2} = \frac{1}{2}\sqrt[3]{12}$$

Chapter 32

Resolving Radicals

If a radicand has a factor that is a perfect square:

In the case of square roots, express the radical as a product

of the perfect square times the remaining factor.

$$\sqrt{20} = \sqrt{5 * 4} = \sqrt{4} * \sqrt{5} = 2\sqrt{5}$$

$$\sqrt[3]{16} = \sqrt[3]{8 * 2} = \sqrt[3]{8} * \sqrt[3]{2} = 2\sqrt[3]{2}$$

32.1   Simplify a Radical Expression Example 2

$Simplify$ $5 * (\sqrt{8} - \sqrt{32})$
$5(\sqrt{8} - \sqrt{32})$
$\sqrt{8}$ and $\sqrt{32}$ $\sqrt{5} * (\sqrt{8} - \sqrt{32})$
$5 \left(\sqrt{(4)(2)} - \sqrt{(16)(2)} \right)$ $5 \left(2\sqrt{2} - 4\sqrt{2} \right) = 10\sqrt{2} - 20\sqrt{2} = -10\sqrt{2}$

32.2   Simplify a Radical Expression Example 3

$$\frac{5}{\sqrt{3}} = \frac{5}{\sqrt{3}} * \frac{\sqrt{3}}{\sqrt{3}} = \frac{5\sqrt{3}}{3} = \frac{5}{3}\sqrt{3}$$

32.3   Unity Factor Explanation

$\frac{\sqrt{3}}{\sqrt{3}}$ is a unity factor and multiplying by a unity factor does not change the value of the expression, it only produces an equivalent form of the expression.

$\frac{5}{\sqrt{3}}$ is rationalized by multiplying the fraction by $\frac{\sqrt{3}}{\sqrt{3}}$ equals 1, a unity multiplier $\frac{5}{\sqrt{3}}$ This yields $5\sqrt{3}$and 3 in the denominator which satisfies the requirement that In a solved radical expression the denominator cannot contain a radical expression,
also $\frac{\sqrt{2}}{\sqrt{2}}$ is a unity factor.

32.4   Simplify a Cubics Example 4

Using the same rules as the previous example, Multiply the fraction by a unity factor that will make the denominator a perfect cube

Solve

$$\frac{\sqrt[3]{3}}{\sqrt[3]{2}}$$

Multiply

$$\frac{\sqrt[3]{3}}{\sqrt[3]{2}}$$

by

$$\frac{\sqrt[3]{2^2}}{\sqrt[3]{2^2}}$$

and take the cube root of the denominator to simplify. We can also express the result as

$$\frac{\sqrt[3]{3*2^2}}{\sqrt[3]{2*2^2}} = \frac{\sqrt[3]{3*2^2}}{\sqrt[3]{2^3}} = \frac{\sqrt[3]{12}}{2} = \frac{1}{2}\sqrt[3]{12}$$

32.5 Unity Factor

The $[\frac{\sqrt{3}}{\sqrt{3}}]$ is rationalized by multiplying the fraction by $[\frac{\sqrt{3}}{\sqrt{3}}]$ equals 1, a unity multiplier $\frac{5}{\sqrt{3}}$. This yields $\frac{5\sqrt{3}}{3}$ which satisfies the requirement that, In a solved radical expression the denominator cannot contain a radical. This can be re-wrtten as $\frac{1}{3} * 5\sqrt{3}$

32.6 Rationalizing the denominator

$$\sqrt{\frac{2x^2y}{3}} = \left(\sqrt{\frac{2x^2y}{3}} * \frac{\sqrt{3}}{\sqrt{3}}\right) = \frac{\sqrt{6x^2y}}{3}$$

$$\frac{\sqrt{x^2}\sqrt{6y}}{3} = \frac{x\sqrt{6y}}{3}$$

32.7 Perfect squares reduction Example 6

$$\sqrt{\frac{169}{9}} = \frac{13}{3}$$

Since both numerators are perfect squares take the square root of both terms.

Chapter 33

Multiplying-and-dividing-roots

33.1   Product rule for radicals

$$\sqrt{x} \bullet \sqrt{y} = \sqrt{xy}$$

$$\sqrt{3} \bullet \sqrt{2} = \sqrt{6}$$

$$\sqrt{5} \bullet \sqrt{x} = \sqrt{5x}$$

Solve the following:

$$\left(\sqrt{3} + 2\sqrt{5}\right)\left(\sqrt{3} - 4\sqrt{5}\right)$$

Solve by using FOIL (First, Outer, Inner, Last)

$$
\begin{array}{cccc}
\textit{First} & \textit{Outer} & \textit{Inner} & \textit{Last} \\
\sqrt{3} \bullet \sqrt{3} & \sqrt{3} \bullet \left(-4\sqrt{5}\right) & 2\sqrt{5} \bullet \sqrt{3} & 2\sqrt{5} \bullet \left(-4\sqrt{5}\right) \\
\\
3 & -4\sqrt{15} \quad + & 2\sqrt{15} & -8\,(5)
\end{array}
$$

Combined middle terms

$$-37 - 2\sqrt{15} \qquad \text{Answer}$$

33.2 Special products rule application

The special products rule for an algebraic term,squared.

$$(a + b)^2 = a^2 + 2ab + b^2$$
$$(a - b)^2 = a^2 - 2ab + b^2$$

Apply the special products rule to solve $\left(\sqrt{10} - 7\right)^2$

$$\left(\sqrt{10} - 7\right)^2; \quad \text{Let } a = \sqrt{10} \quad \text{and} \quad b = 7$$

$$\left(\sqrt{10} - 7\right)^2 = \left(\sqrt{10}\right)^2 - 2 \bullet \sqrt{10} \bullet 7 + (7)^2$$

$$10 - 14\sqrt{10} + 49$$
$$59 - 14\sqrt{10} \quad \text{answer}$$

33.3 Reducing binomial multiplication with radicals

The following is an example reducing binomial multiplication radicals using FOIL method. Solve the following:

$$\left(\sqrt{10} - \sqrt{5}\right) \bullet \left(\sqrt{5} + \sqrt{20}\right)$$

General form:

$$(a - b) \bullet (b + c)$$

$$(a - b) \bullet (b + c) = ab + bc - b^2 - bc$$

$$\left(\sqrt{10} \bullet \sqrt{5}\right) + \left(\sqrt{10} \bullet \sqrt{20}\right) - \left(\sqrt{5}\right)^2 - \left(\sqrt{5} \bullet \sqrt{20}\right)$$

$$\sqrt{50} + \sqrt{200} - 5 - \sqrt{100}$$

Reduce the radicals by factoring out the perfect squares.

$$\sqrt{25 \bullet 2} + \sqrt{100 \bullet 2} - 5 - \sqrt{100}$$

$$5\sqrt{2} + 10\sqrt{2} - 5 - 10$$

Collect like terms.

$$15\sqrt{2} - 15 \quad \text{answer}$$

33.4 Complete the solution by rationalizing the denominator

$$\sqrt{\frac{7}{13}} \bullet \sqrt{\frac{13}{3}} = \sqrt{\frac{7 \bullet 13}{13 \bullet 3}} = \sqrt{\frac{7}{3}}$$

Rationalize the denominator by multiplying by

$$\sqrt{\frac{3}{3}}$$

$$= \sqrt{\frac{7}{3}} \bullet \sqrt{\frac{3}{3}} = \sqrt{\frac{7 \bullet 3}{9}} = \frac{\sqrt{21}}{3} \quad \text{or} \quad \frac{1}{3}\sqrt{21} \text{ answer}$$

Appendix A

Review of Quadratics

A.1 Quadratic equation real solution

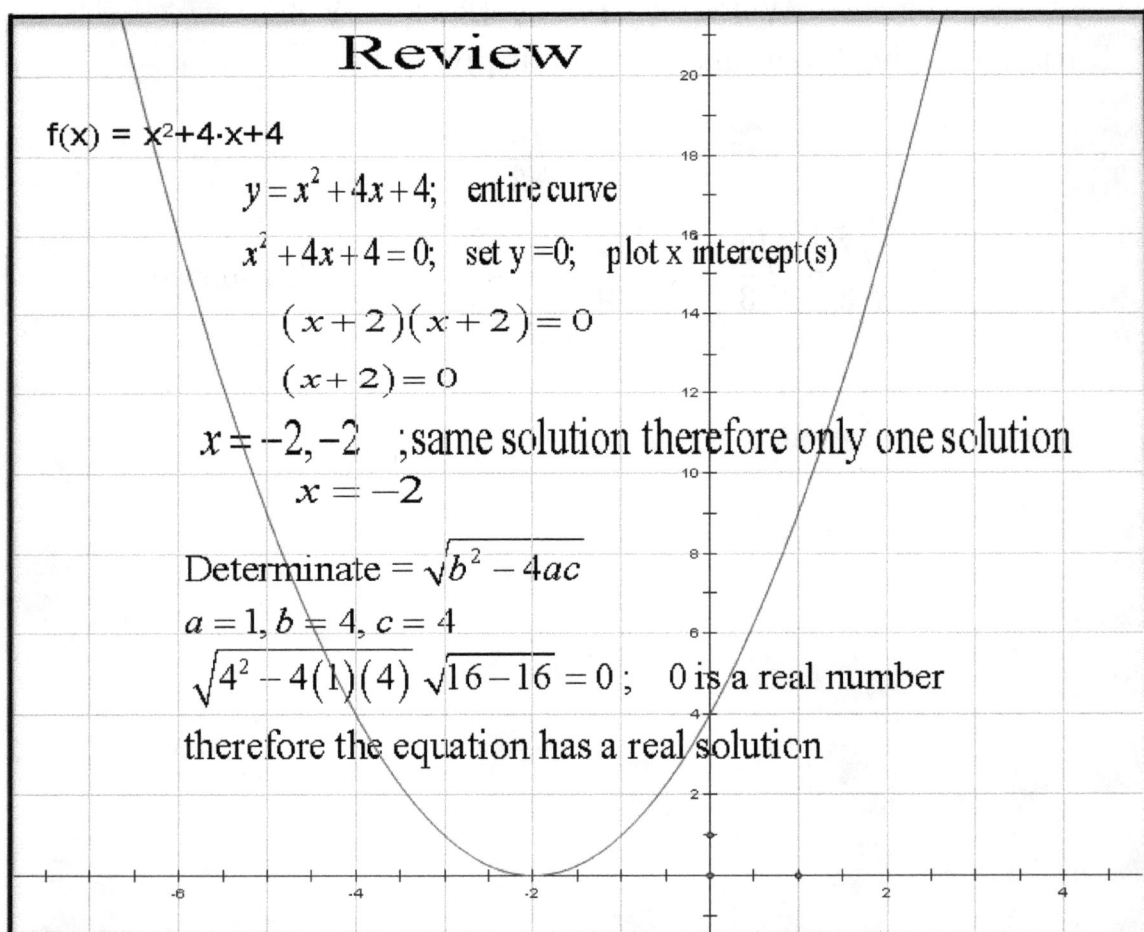

Review

f(x) = x²+4·x+4

$y = x^2 + 4x + 4;$ entire curve

$x^2 + 4x + 4 = 0;$ set y = 0; plot x intercept(s)

$(x+2)(x+2) = 0$

$(x+2) = 0$

$x = -2, -2$;same solution therefore only one solution

$x = -2$

Determinate $= \sqrt{b^2 - 4ac}$

$a = 1, b = 4, c = 4$

$\sqrt{4^2 - 4(1)(4)} \sqrt{16 - 16} = 0;$ 0 is a real number

therefore the equation has a real solution

A.2   Quadratic equation Imaginary solution

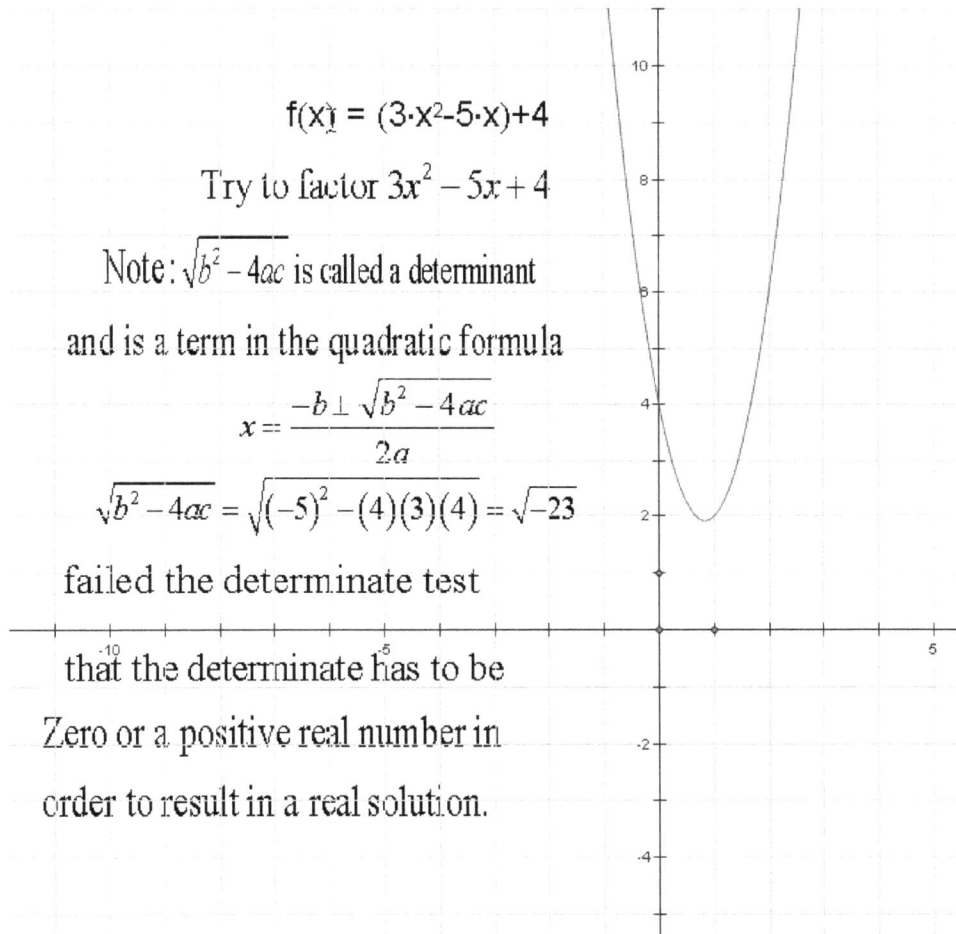

f(x) = (3·x²-5·x)+4

Try to factor $3x^2 - 5x + 4$

Note: $\sqrt{b^2 - 4ac}$ is called a determinant

and is a term in the quadratic formula

$$x = \frac{-b \pm \sqrt{b^2 - 4ac}}{2a}$$

$$\sqrt{b^2 - 4ac} = \sqrt{(-5)^2 - (4)(3)(4)} = \sqrt{-23}$$

failed the determinate test

that the determinate has to be

Zero or a positive real number in

order to result in a real solution.

Appendix B

Distance-Formulas

B.1 Physics Problem

B.2 Formula for Vertical distance

This formula is used in a vertical problem found on page 99

Vertical Distance: $= V_{y0}t - \frac{1}{2}gt^2$

Acceleration due to gravity $g = 32\frac{ft}{\sec^2}$

$y = f(t)\mathrm{y}$

is dependent on t for its solution

Change the order of terms

to the format of a quadratic equation

Replace y with f(t).

Note: f(t) states that y

is dependent on t for its solution

Original equation: $y = V_{y0}t - \frac{1}{2}gt^2$

$H = s(t)$ equals height of ball

at any time, and is dependent on

the time elapsed.

$H = s(t) = -\frac{1}{2}gt^2 + V_{y0}t \ + \ H_0$ feet

Where H_0 = initial height above ground

Acceleration due to gravity $32\frac{ft}{sec^2}$

This fits into the general format

for a Quadratic Equation

$y = f\left(x\right) = ax^2 + bx + c$

$s\left(t\right) = -16t^2 + 16t + 192$

$y = f\left(x\right) = -\frac{1}{2}\left(32\frac{ft}{sec^2}\right)t^2 + V_{y0}t \ + \ 0$ feet

$y = f\left(x\right) = -16\ t^2 + V_{y0}t \ + \ 0$ feet

$f\left(x\right) = -16\ t^2 + V_{y0}t \ + \ 0$ feet

In the example given earlier

$V_{y0}t$ the initial velocity was $16\frac{ft}{sec}$

The initial Height was 192 feet

Substituting these values

$s\left(t\right) = -16t^2 + 16t + 192$

$s(t) = -\frac{1}{2}gt^2 + V_{y0}t \ + \ x$ feet

Thrown ball - Quadratic Equation

t = 3 seconds to reach 96 feet

t = 4 seconds to reach 0 feet

Where 0 feet is at ground level.

B.3 Vertical height vs time

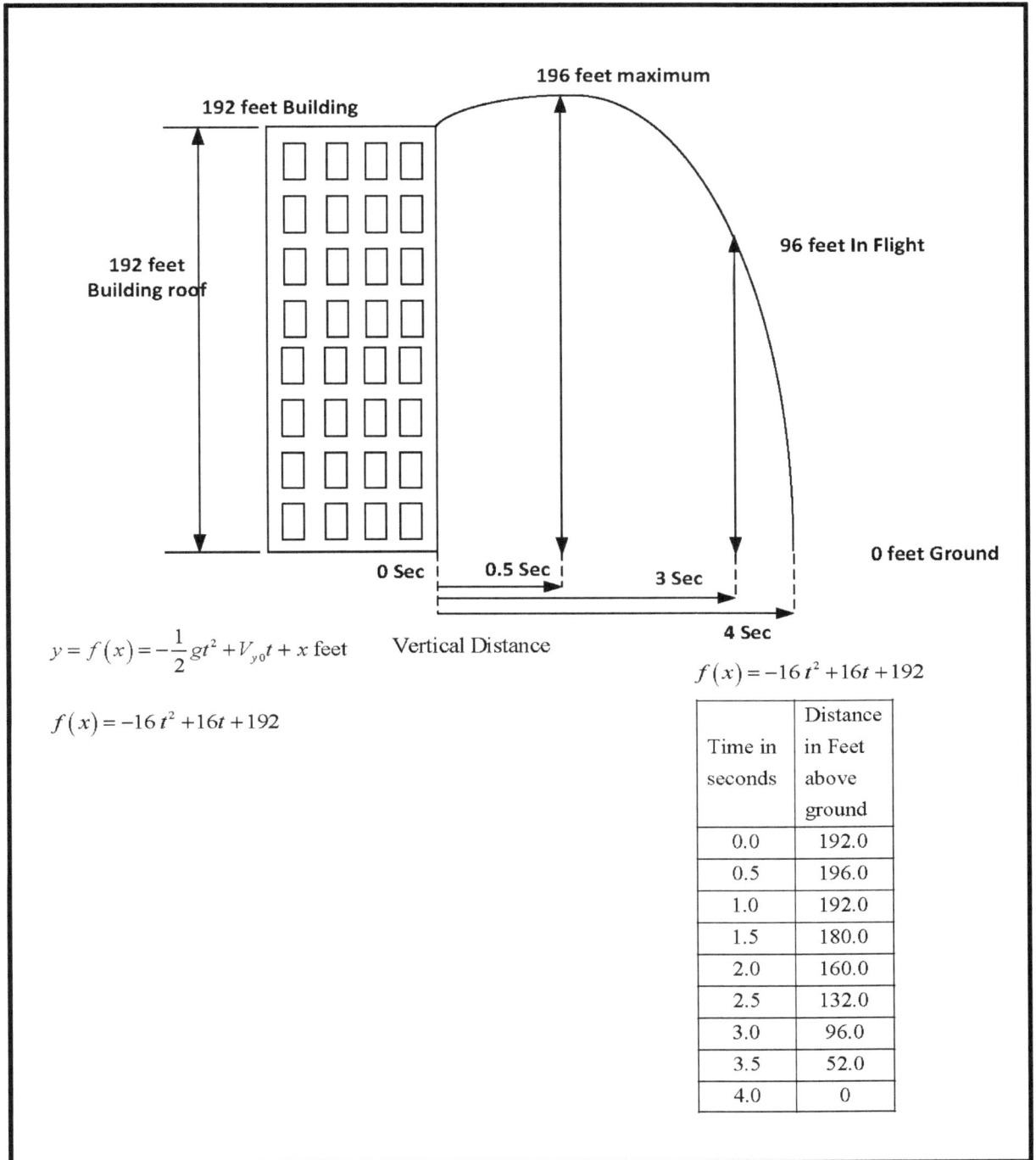

196 feet maximum

192 feet Building

96 feet In Flight

192 feet Building roof

0 feet Ground

0 Sec **0.5 Sec** **3 Sec** **4 Sec**

$$y = f(x) = -\frac{1}{2}gt^2 + V_{y0}t + x \text{ feet}$$

Vertical Distance

$$f(x) = -16\,t^2 + 16t + 192$$

$$f(x) = -16\,t^2 + 16t + 192$$

Time in seconds	Distance in Feet above ground
0.0	192.0
0.5	196.0
1.0	192.0
1.5	180.0
2.0	160.0
2.5	132.0
3.0	96.0
3.5	52.0
4.0	0

Appendix C

Approximate numbers

C.1 Number description

- Numbers can be Integers or Approximate Numbers
- Integers are used in counting.
- In Engineering Approximate numbers are important.
- Every measurement is approximate and depends on the accuracy of the instrument taking the measurement.
- This is true in any type of science.

C.2 Critical factors of approximate numbers

- Precision
- Accuracy

C.3 Definition approximate numbers

- Precision depends on number of places after the decimal point.
- Accuracy depends on number of significant digits. (see chart)

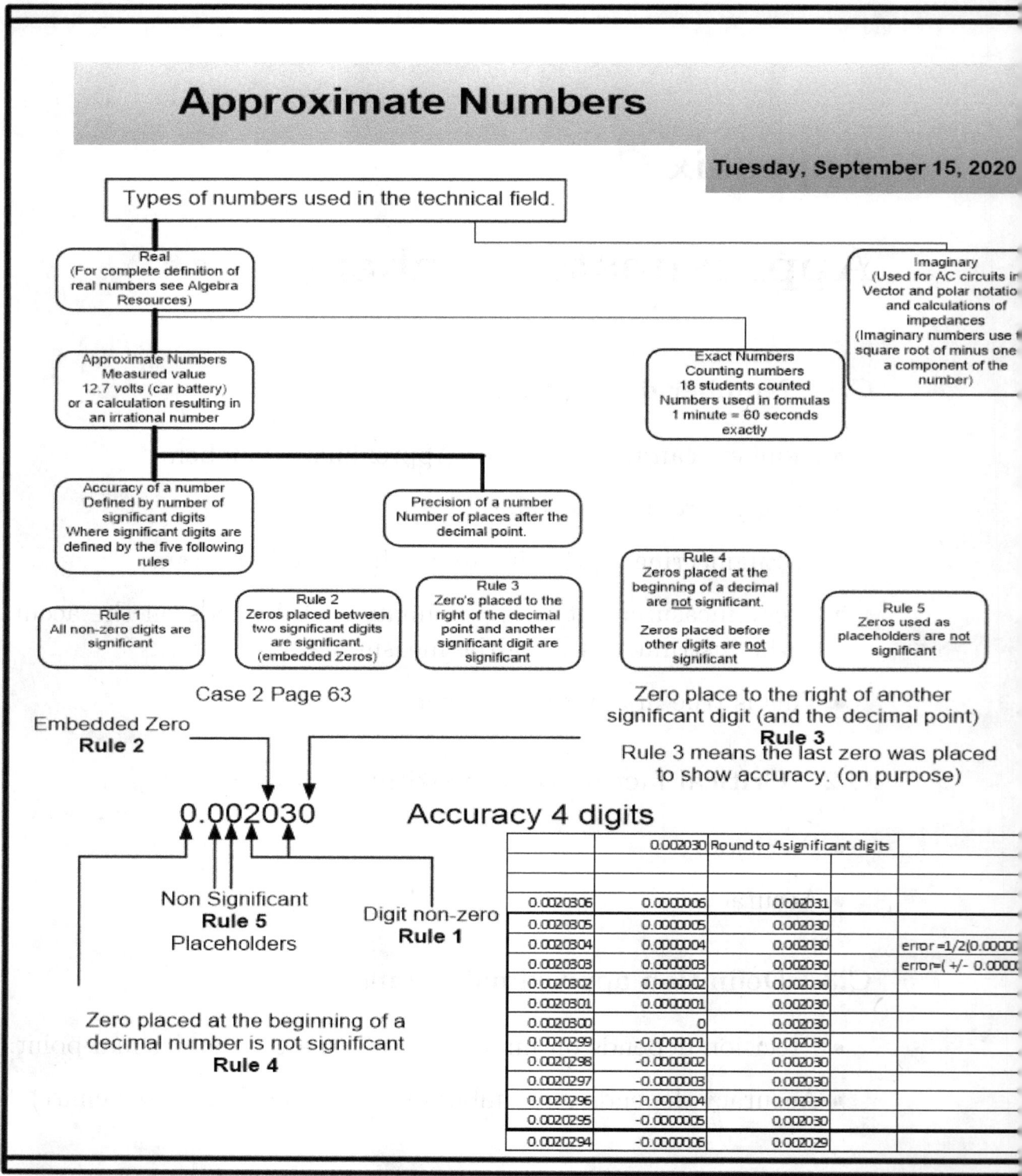

Figure C.1: Approximation of numbers

Appendix D

Algebra rules apples other higher-level math

D.1 Trigonometry Discussion

In trig we layer on rules of trigonometric functions also.

In Trigonometry, functions of trigonometry replace some of the variables but the rules for manipulating variables are the same .

i.e. We use $\sin(x)$ $\cos(x)$ or $\tan(x)$ as variables as well as x and y and z when needed to solve the problem.

D.2 Calculus Discussion

In calculus the some of the variables are constantly changing according the rules of calculus.

Variables are time, speed, distance and acceleration,

The variables would be t,s,v,a, in acceleration problems.

Some of the variables will continually change during the duration of the problem

Appendix E

Mathematics in Electrical Eng

Algebra is used in Mechanical Engineering,Civil Engineering, Chemistry and all other sciences.

E.1   Electrical Engineering

Both in Electrical-Engineering-and-Using-

Practical-Math-Applications

Electrical Engineering and Practical

Math Applications used in solving Electrical Design Problems.

Were used every day of my projects. It was our task to debug new

designs and design test equipment and make both the designs and test

equipment work before the design could be approved.

E.2 Proofs as Student handouts to add to explanation of principles.

We used a lot of practical mathematics putting to use the Math and physics and electrical theory that we learned in engineering school. My teaching was to help understand the theory to make tests and later career problem solving easier and more fun. To that end I started writing proofs that were used as handouts to the students.

E.3 Proofs

If a student understands how to derive a formula the student can adapt the formula for solving similar math problems.

Note: The imaginary coordinate system is used in electrical work where $-i$ values are capacitance and $+i$ values are inductance and resistance is plotted along the x-axis.

In Electrical Engineering the curve showing a transistor or diode curve also tells us in a v versus vi curve indicates what setting will give us the most stability. This is very important when designing an amplifier or a computer.

In Mechanical Engineering the elasticity the linear portion of the curve show how much stress can be applied to a metal, wood or plastic device. Too much stress will cause a breaking point potentially. This used in designing a new part.

Appendix F

Calculus

F.1   Approximate-vs-Integral

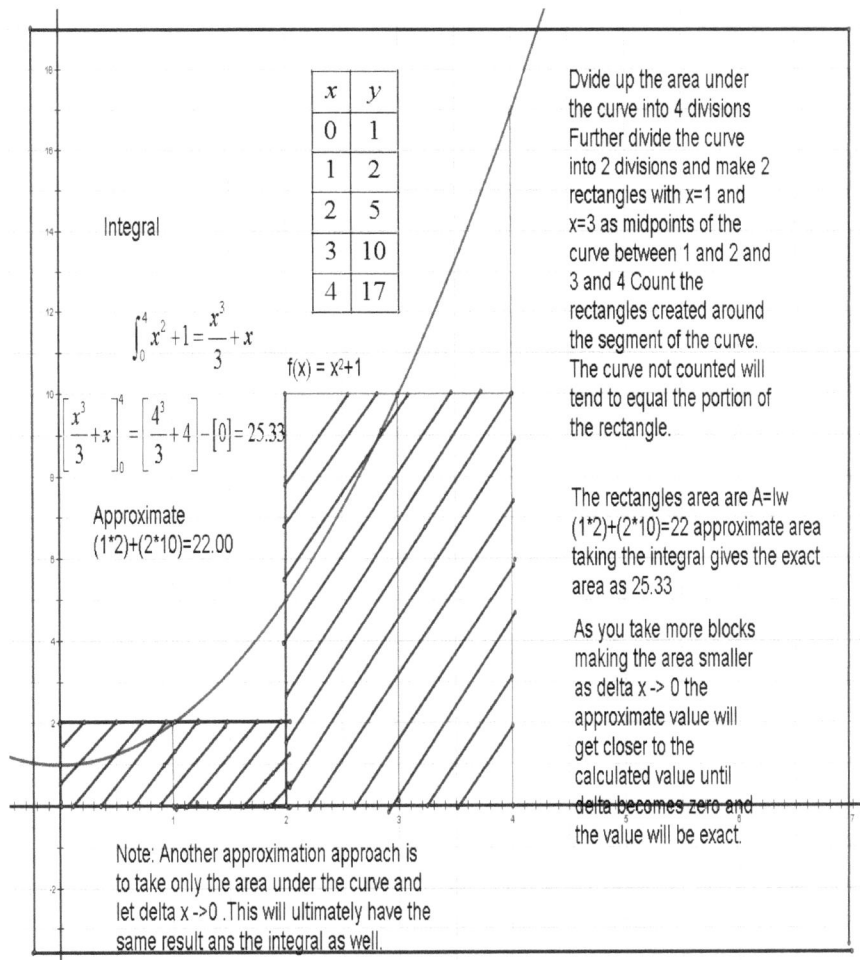

x	y
0	1
1	2
2	5
3	10
4	17

Integral

$$\int_0^4 x^2 + 1 = \frac{x^3}{3} + x$$

$$\left[\frac{x^3}{3} + x\right]_0^4 = \left[\frac{4^3}{3} + 4\right] - [0] = 25.33$$

$f(x) = x^2 + 1$

Approximate
(1*2)+(2*10)=22.00

Dvide up the area under the curve into 4 divisions Further divide the curve into 2 divisions and make 2 rectangles with x=1 and x=3 as midpoints of the curve between 1 and 2 and 3 and 4 Count the rectangles created around the segment of the curve. The curve not counted will tend to equal the portion of the rectangle.

The rectangles area are A=lw (1*2)+(2*10)=22 approximate area taking the integral gives the exact area as 25.33

As you take more blocks making the area smaller as delta x -> 0 the approximate value will get closer to the calculated value until delta becomes zero and the value will be exact.

Note: Another approximation approach is to take only the area under the curve and let delta x ->0 .This will ultimately have the same result ans the integral as well.

171

F.2   secant

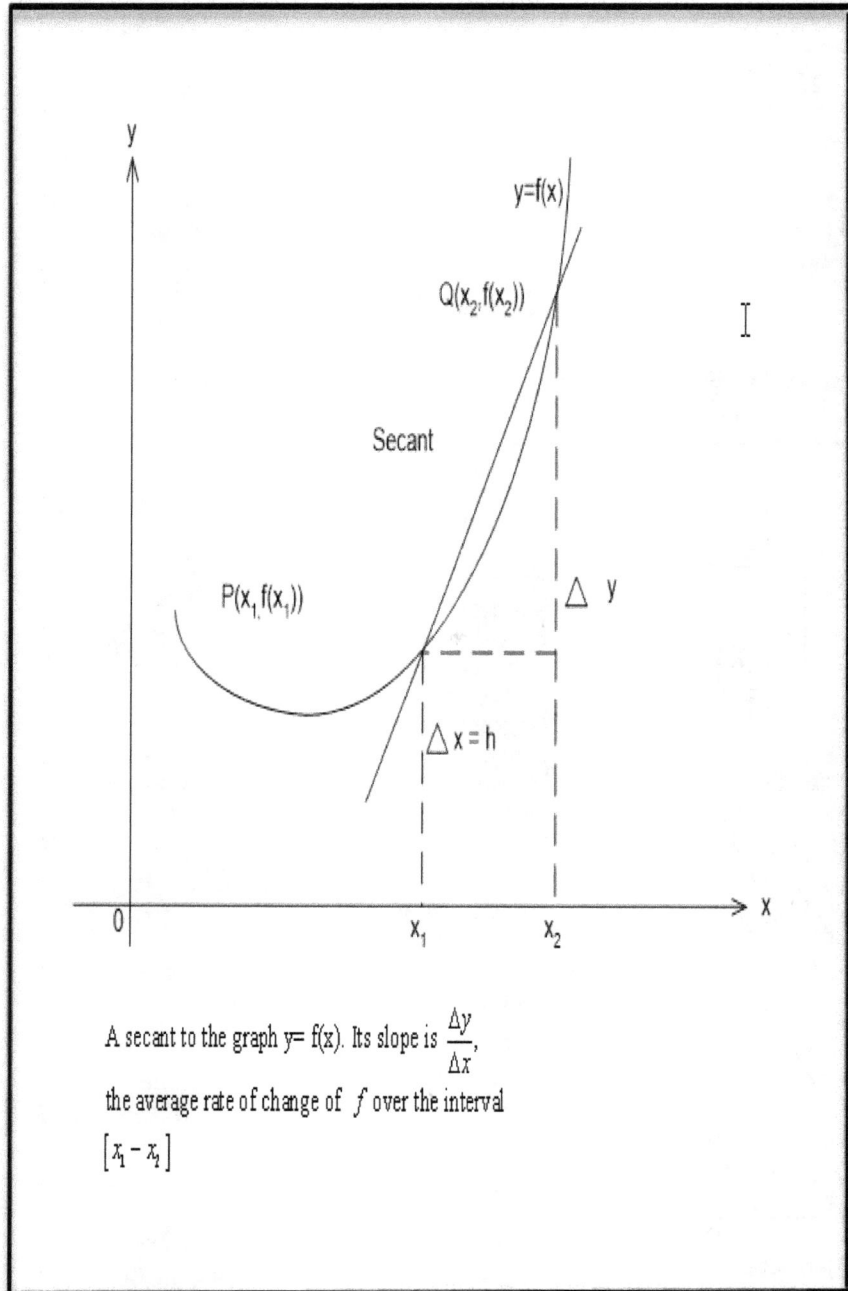

A secant to the graph y= f(x). Its slope is $\dfrac{\Delta y}{\Delta x}$, the average rate of change of f over the interval $[x_1 - x_2]$

Figure F.1: secant

F.3 Limit of delta x

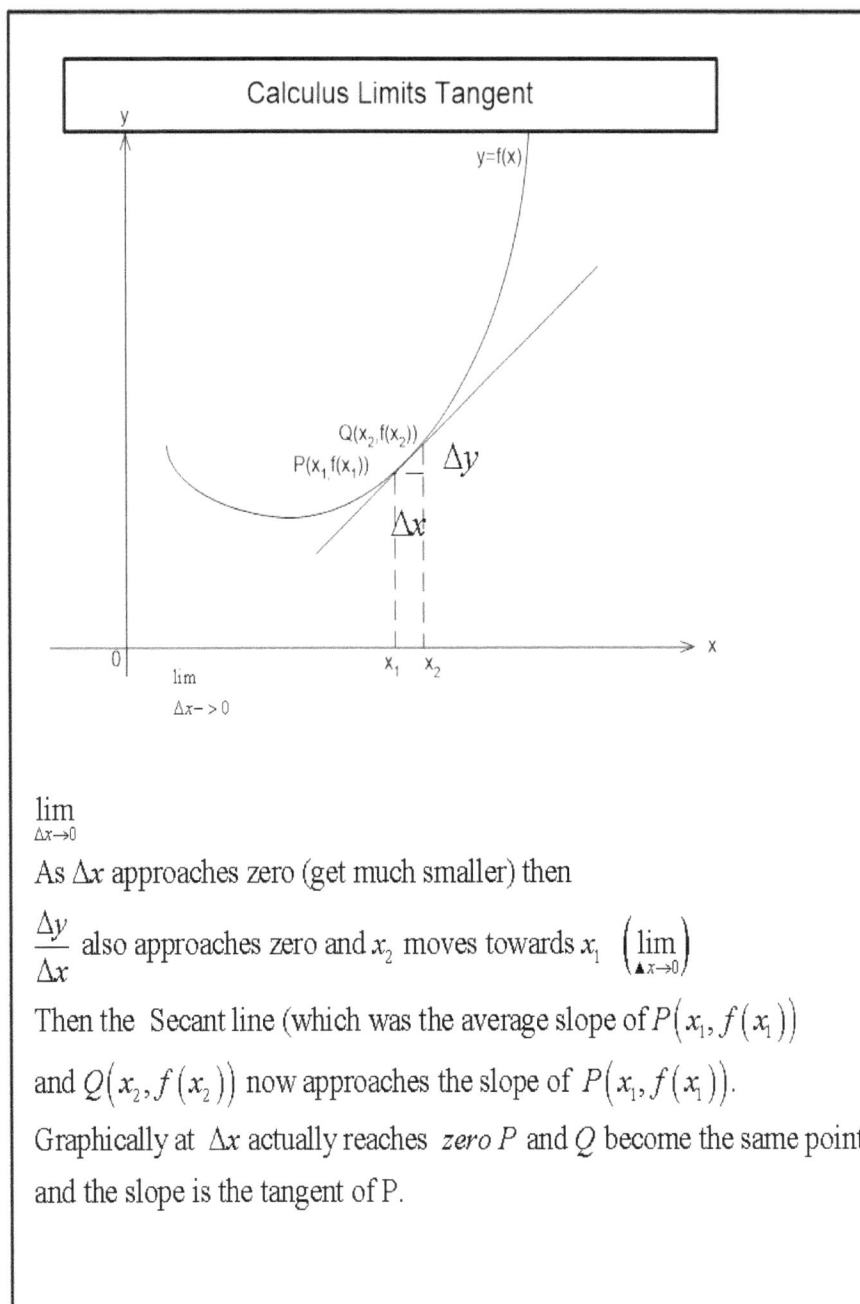

$$\lim_{\Delta x \to 0}$$

As Δx approaches zero (get much smaller) then

$\dfrac{\Delta y}{\Delta x}$ also approaches zero and x_2 moves towards x_1 $\left(\lim_{\blacktriangle x \to 0}\right)$

Then the Secant line (which was the average slope of $P\left(x_1, f\left(x_1\right)\right)$

and $Q\left(x_2, f\left(x_2\right)\right)$ now approaches the slope of $P\left(x_1, f\left(x_1\right)\right)$.

Graphically at Δx actually reaches *zero* P and Q become the same point

and the slope is the tangent of P.

Figure F.2: Calculus Limits Tangent

Appendix G

Math used in Physics

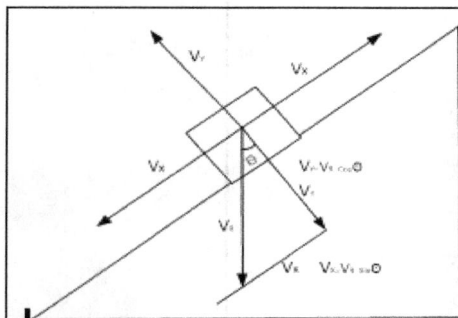

This outline shows the Physics Force Vectors

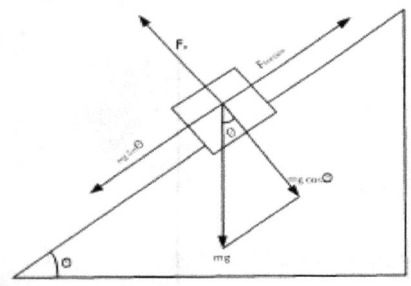

Trigonometry

Geometry

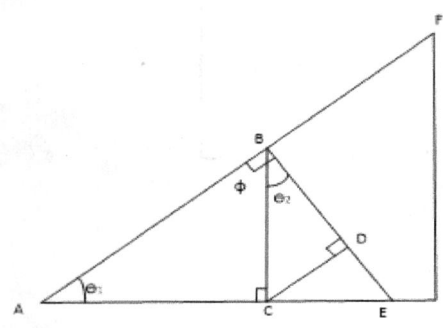

Force Vector Diagram

Forces of a block on a ramp. Where a vector has magnitude and direction. Newton's 3rd law For every action there is an equal and opposite reaction. The interaction of the block on the ramp creates a set of interactive forces in both the vertical and horizontal direction. The block also has a force of gravity in the y direction

Trigonometry

$$\sin\theta = \frac{\text{Opp}}{\text{Hyp}} = \frac{F_{gx}}{F}; therefore\ F_{gx} = F\sin\theta$$

$$\cos\theta = \frac{\text{Adj}}{\text{Hyp}} = \frac{F_{gy}}{F}; therefore\ F_{gy} = F\cos\theta$$

Geometry

Prove that the ramp angle CAB is congruent to angle CBD.
AB $\perp$ BD and CD $\perp$ BD and AC $\perp$ BC therefore forming a series of right angle triangles namely ▲ABC ▲CBD.

Where CAB= Θ_1 and CBE= Θ_2 and ABC= Φ.

1) $\Theta_1 + \Phi = 90^0$ therefore $\Theta_1 = 90^0 - \Phi$
2) $\Theta_2 + \Phi = 90^0$ therefore $\Theta_2 = 90^0 - \Phi$
   Statmt 1) sum of a rt $\Delta = 180^0$
   Statmt 2) sum complementary angles equals 90^0
3) Then $\Theta_1 = \Theta_2$ and are congruent

174

Appendix H

convert-units.png

Question:6 yds = ? inches (")

From table 2.3   1 yd = 3 ft

From table 2.3 $1 \text{ in} = \dfrac{1}{12} \text{ ft}$

Convert the equivalent unity conversion units

$$Unity\ 1 = \frac{1\,in}{\dfrac{1}{12}\,ft} = \left[\frac{1\,in}{\dfrac{1}{12}\,ft} \bullet \frac{12}{12}\right] = \frac{12\,in}{1\,ft} \quad or \quad \frac{1\,ft}{12\,in}$$

$$Unity\ 1 = \frac{1\,yd}{3\,ft} \quad also \quad \frac{3\,ft}{1\,yd}$$

Since the numerator and the denominator are the same they are interchangeable.

Question:6 yds = ? inches (")

$$\frac{6\,yds}{1} \bullet \frac{3\,ft}{1\,yd} \bullet \frac{12\,in}{1\,ft} = \frac{6}{1} \bullet \frac{3}{1} \bullet \frac{12\,in}{1} = 216\,in = 216"$$

Yard to Inch
converter

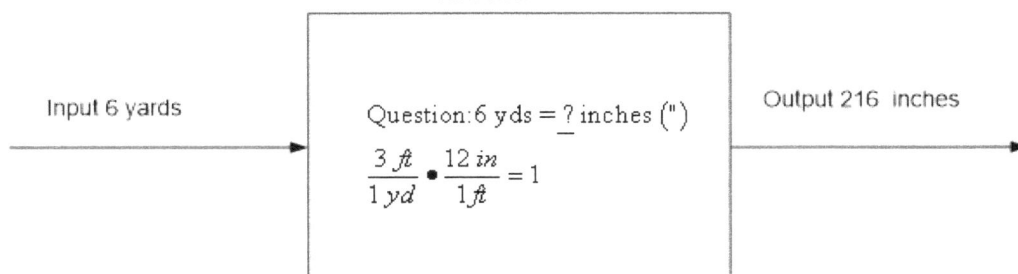

Input 6 yards   Question:6 yds = ? inches (")   Output 216 inches

$$\frac{3\,ft}{1\,yd} \bullet \frac{12\,in}{1\,ft} = 1$$

Question:6 yds = ? inches (")

$$\frac{6\,yds}{1} \bullet \frac{3\,ft}{1\,yd} \bullet \frac{12\,in}{1\,ft} = \frac{6}{1} \bullet \frac{3}{1} \bullet \frac{12\,in}{1} = 216\,in = 216"$$

Appendix I

Calculus

I.1 Calculus:Approximate area

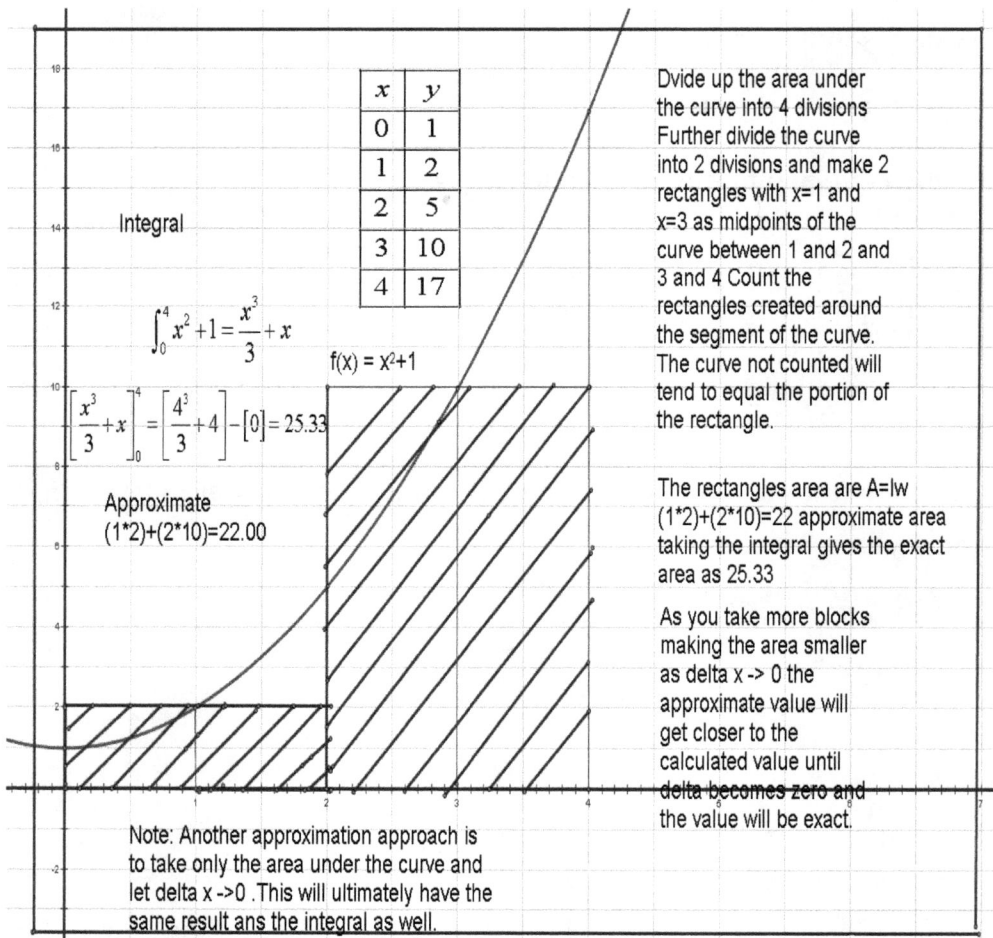

x	y
0	1
1	2
2	5
3	10
4	17

Integral

$$\int_0^4 x^2+1 = \frac{x^3}{3}+x$$

$$\left[\frac{x^3}{3}+x\right]_0^4 = \left[\frac{4^3}{3}+4\right]-[0] = 25.33$$

$f(x) = x^2+1$

Approximate
(1*2)+(2*10)=22.00

Dvide up the area under the curve into 4 divisions Further divide the curve into 2 divisions and make 2 rectangles with x=1 and x=3 as midpoints of the curve between 1 and 2 and 3 and 4 Count the rectangles created around the segment of the curve. The curve not counted will tend to equal the portion of the rectangle.

The rectangles area are A=lw (1*2)+(2*10)=22 approximate area taking the integral gives the exact area as 25.33

As you take more blocks making the area smaller as delta x -> 0 the approximate value will get closer to the calculated value until delta becomes zero and the value will be exact.

Note: Another approximation approach is to take only the area under the curve and let delta x ->0 .This will ultimately have the same result ans the integral as well.

I.2 secant

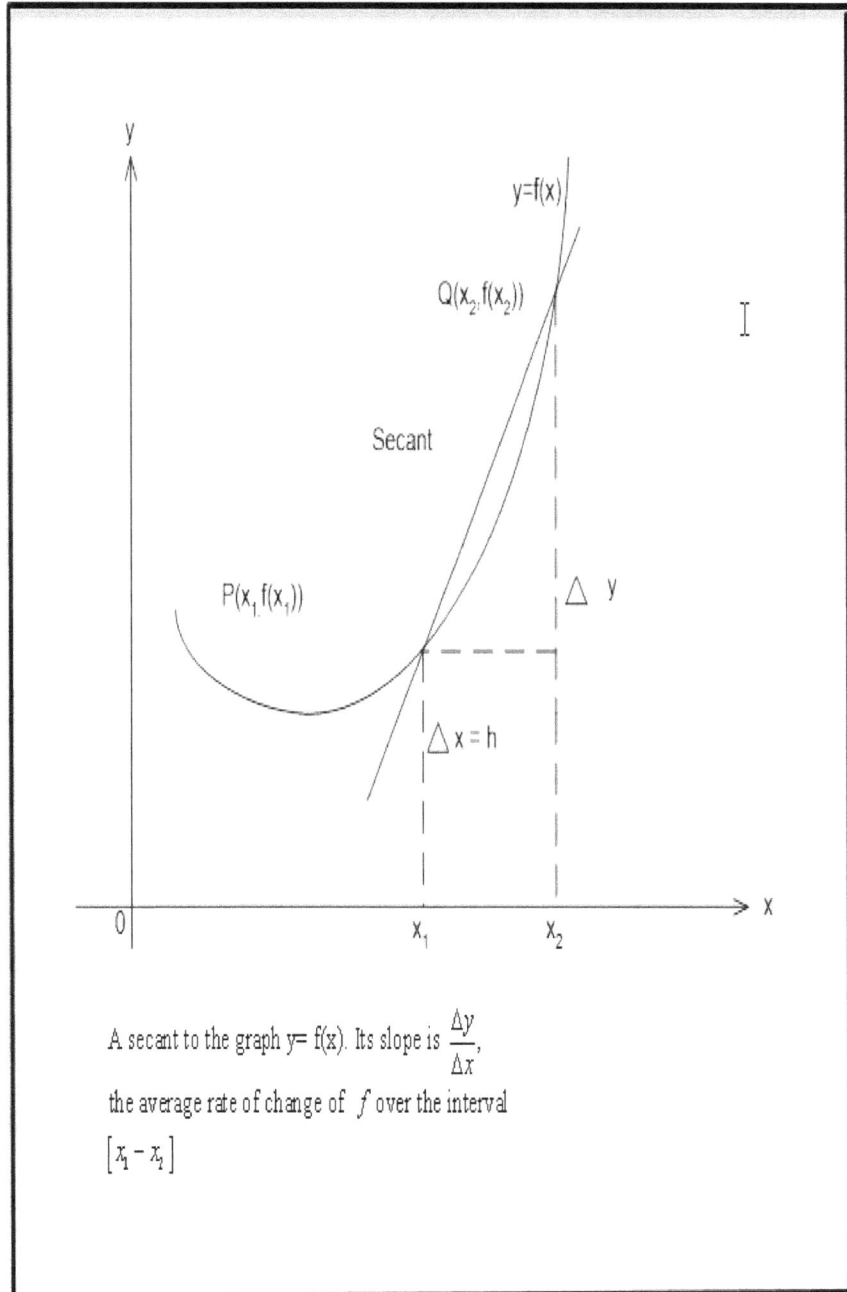

A secant to the graph y= f(x). Its slope is $\frac{\Delta y}{\Delta x}$, the average rate of change of f over the interval $[x_1 - x_1]$

Figure I.1: secant

I.3   Limit of delta x

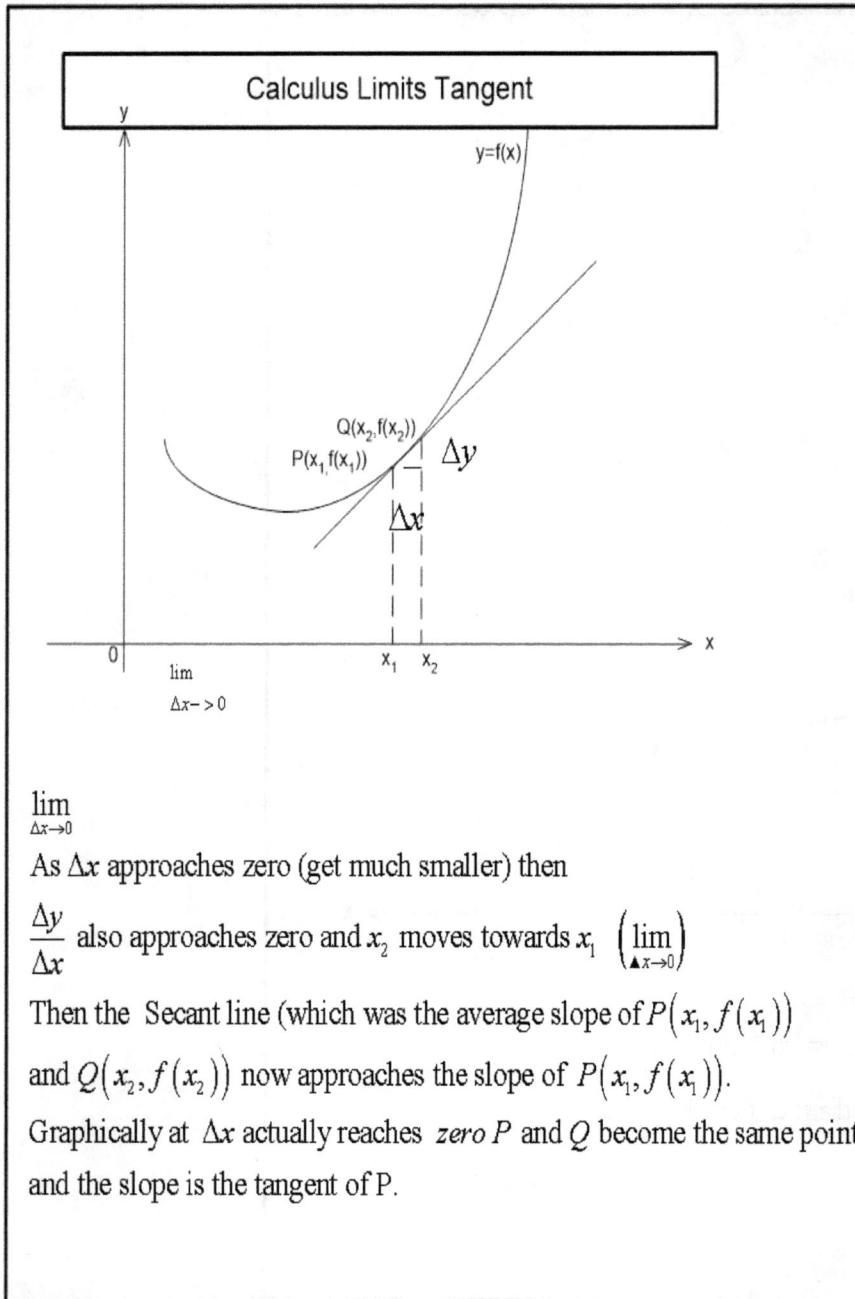

Figure I.2: Calculus Limits Tangent

I.4   finite-approximation.png

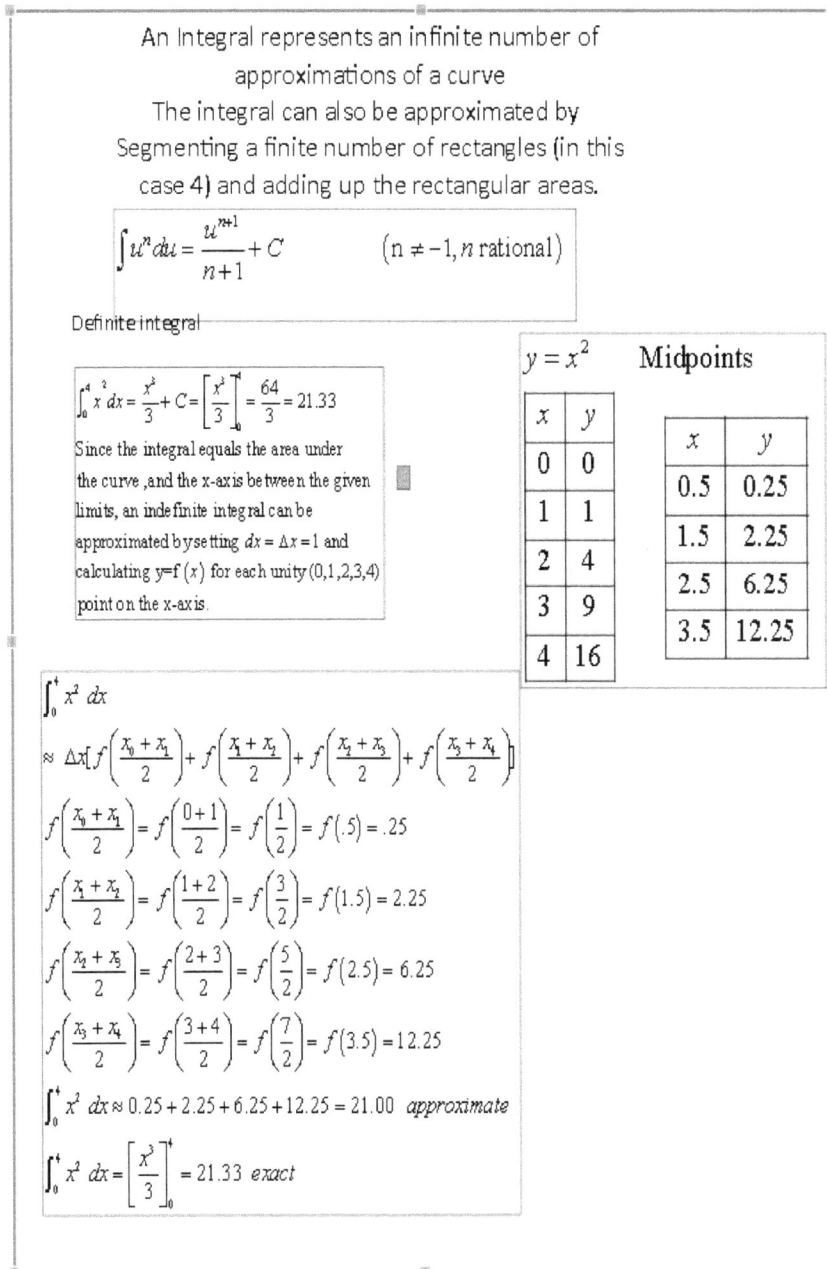

An Integral represents an infinite number of
approximations of a curve
The integral can also be approximated by
Segmenting a finite number of rectangles (in this
case 4) and adding up the rectangular areas.

$$\int u^n du = \frac{u^{n+1}}{n+1} + C \qquad \left(n \neq -1, n \text{ rational}\right)$$

Definite integral

$$\int_0^4 x^2 \, dx = \frac{x^3}{3} + C = \left[\frac{x^3}{3}\right]_0^4 = \frac{64}{3} = 21.33$$

Since the integral equals the area under
the curve, and the x-axis between the given
limits, an indefinite integral can be
approximated by setting $dx = \Delta x = 1$ and
calculating $y = f(x)$ for each unity $(0,1,2,3,4)$
point on the x-axis.

$y = x^2$ Midpoints

x	y
0	0
1	1
2	4
3	9
4	16

x	y
0.5	0.25
1.5	2.25
2.5	6.25
3.5	12.25

$$\int_0^4 x^2 \, dx$$

$$\approx \Delta x \left[f\left(\frac{x_0 + x_1}{2}\right) + f\left(\frac{x_1 + x_2}{2}\right) + f\left(\frac{x_2 + x_3}{2}\right) + f\left(\frac{x_3 + x_4}{2}\right) \right]$$

$$f\left(\frac{x_0 + x_1}{2}\right) = f\left(\frac{0+1}{2}\right) = f\left(\frac{1}{2}\right) = f(.5) = .25$$

$$f\left(\frac{x_1 + x_2}{2}\right) = f\left(\frac{1+2}{2}\right) = f\left(\frac{3}{2}\right) = f(1.5) = 2.25$$

$$f\left(\frac{x_2 + x_3}{2}\right) = f\left(\frac{2+3}{2}\right) = f\left(\frac{5}{2}\right) = f(2.5) = 6.25$$

$$f\left(\frac{x_3 + x_4}{2}\right) = f\left(\frac{3+4}{2}\right) = f\left(\frac{7}{2}\right) = f(3.5) = 12.25$$

$$\int_0^4 x^2 \, dx \approx 0.25 + 2.25 + 6.25 + 12.25 = 21.00 \quad approximate$$

$$\int_0^4 x^2 \, dx = \left[\frac{x^3}{3}\right]_0^4 = 21.33 \quad exact$$

Figure I.3: finite-approximation.png

I.5   integral-by-parts

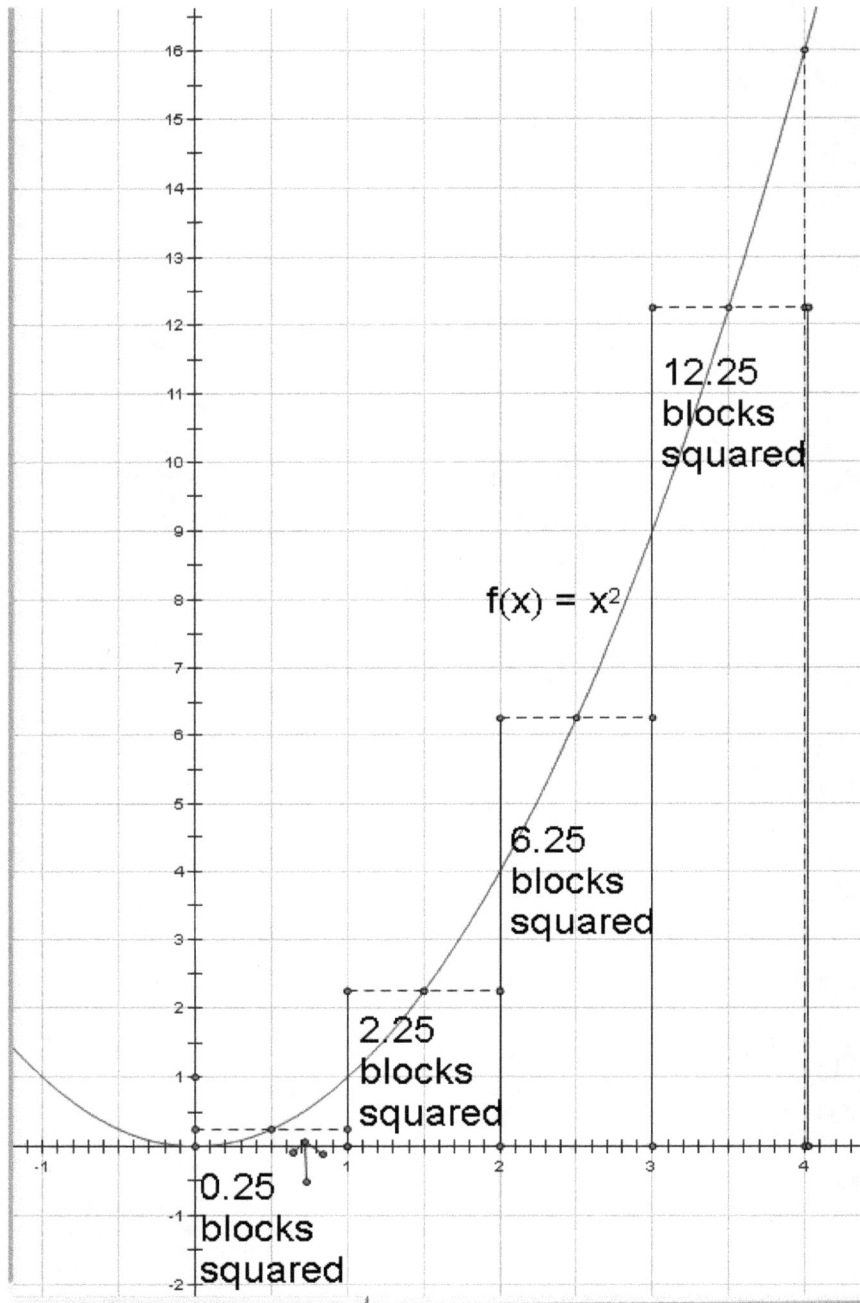

Figure I.4: integral-by-parts